101 KEY IDEAS

ECOLOGY

ECOLOGY

Paul Mitchell

LLYSFASI COLLEGE	
C00012353	
Cypher	20.7.01
577	£4.99
8756	

TEACH YOURSELF BOOKS

For UK orders: please contact Bookpoint Ltd, 78 Milton Park, Abingdon, Oxon OX14 4TD. Telephone: (44) 01235 400414, Fax: (44) 01235 400454. Lines are open from 9.00–6.00, Monday to Saturday, with a 24 hour message answering service. Email address: orders@bookpoint.co.uk

For USA & Canada orders: please contact NTC/Contemporary Publishing, 4255 West Touhy Avenue, Lincolnwood, Illinois 60646–1975, U.S.A. Telephone: (847) 679 5500, Fax: (847) 679 2494.

Long renowned as the authoritative source for self-guided learning – with more than 30 million copies sold worldwide – the *Teach Yourself* series includes over 200 titles in the fields of languages, crafts, hobbies, business and education.

British Library Cataloguing in Publication Data
A catalogue record for this title is available from The British Library.

Library of Congress Catalog Card Number: On file

First published in UK 2000 by Hodder Headline Plc, 338 Euston Road, London, NW1 3BH.

First published in US 2000 by NTC/Contemporary Publishing, 4255 West Touhy Avenue, Lincolnwood (Chicago), Illinois 60646–1975 USA.

The 'Teach Yourself' name and logo are registered trade marks of Hodder & Stoughton Ltd.

Cover design and illustration by Mike Stones.

Typeset by Transet Limited, Coventry, England.
Printed in Great Britain for Hodder & Stoughton Educational, a division of Hodder Headline Plc, 338 Euston Road, London NW1 3BH by Cox & Wyman Ltd, Reading, Berkshire.

Impression number   10 9 8 7 6 5 4 3 2 1
Year   2005 2004 2003 2002 2001 2000

Contents

Abyssal zone   1

Balance of nature   2

Behavioural ecology   3

Biocontrol of pests   4

Biodiversity   5

Biogeochemical cycling   6

Biomes   7

Chaos   8

Character displacement   9

Chemical ecology   10

Coevolution   11

Community   12

Community – alternative stable
   states   13

Community assembly   14

Community structure   15

Complexity and stability   16

Coniferous forests (taiga)   17

Conservation   18

Decomposition   19

Density dependence   20

Deserts   21

Disease   22

Dispersal   23

Disturbance   24

Ecological energetics   25

Ecological redundancy   26

Ecology   27

Ecophysiology   28

Ecosystem   29

Ecosystem engineers   30

Ecotoxicology   31

Environment   32

Equilibrium   33

Experimental ecology   34

Fire   35

Food web   36

Functional groups   37

Gaia   38

Generality in ecology   39

Global environmental change   40

Grasslands   41

Guilds   42

Habitat   43

Habitat fragmentation   44

Habitat management   45

Habitat (re)creation   46

Herbivory   47

Historical ecology   48

Indirect effects   49

Interspecific competiton   50

Intraspecific competition   51

Island biogeography theory   52

Keystone species   53

Lakes   54

Landscape ecology   55

Latitudinal diversity gradient   56

Life forms   57

Life history strategies   58

Limiting factors   59

Macroecology   60

Mediterranean shrubland   61

Metapopulation   62

Microbial ecology   63

Microbial loop   64

Minimum viable population   65

Models in ecology   66

Molecular ecology   67

Mutualism   68

Niche   69

Oceans   70

Organisms   71

Parasitism   72

Parasitoids   73

Patchiness   74

Population   75

Population growth   76

Population regulation   77

Predation   78

Predator-mediated
   coexistence   79

Primary production   80

Rarity   81

Resource partitioning   82

Restoration ecology   83

Rivers   84

Savannah   85

Scale in ecology   86

Semantics   87

Species/area relationship   88

Species coexistence   89

Species interactions   90

Species introductions   91

Succession   92

Succession – primary   93

Succession – secondary   94

Symbiosis   95

Temperate forests   96

Top-down/bottom-up   97

Trophic cascade(s)   98

Trophic level(s)   99

Tropical rainforests   100

Tundra   101

Introduction

Welcome to the **Teach Yourself 101 Key Ideas** series. We hope that you will find both this book and others in the series to be useful, interesting and informative. The purpose of the series is to provide an introduction to a wide range of subjects, in a way that is entertaining and easy to absorb.

Each book contains 101 short accounts of key ideas or terms which are regarded as central to that subject. The accounts are presented in alphabetical order for ease of reference. All of the books in the series are written in order to be meaningful whether or not you have previous knowledge of the subject. They will be useful to you whether you are a general reader, are on a pre-university course, or have just started at university.

We have designed the series to be a combination of a text book and a dictionary. We felt that many text books are too long for easy reference, while the entries in dictionaries are often too short to provide sufficient detail. The **Teach Yourself 101 Key Ideas** series gives the best of both worlds! Here are books that you do not have to read cover to cover, or in any set order. Dip into them when you need to know the meaning of a term, and you will find a short, but comprehensive account which will be of real help with those essays and assignments. The terms are described in a straightforward way with a careful selection of academic words thrown in for good measure!

So if you need a quick and inexpensive introduction to a subject, **Teach Yourself 101 Key Ideas** is for you. And incidentally, if you have any suggestions about this book or the series, do let us know. It would be great to hear from you.

Best wishes with your studies!

Paul Oliver
Series Editor

Abyssal zone

The abyssal zone – ocean depths below 2000 metres – occupies over half of the Earth's surface. It is therefore the most typical environment on the planet; yet it remains the least well known. Only with the recent development of deep-sea submarines are we beginning to understand this amazing environment.

The abyssal zone is characterised by constant conditions: it's cold, dark, subject to incredible pressures (over 1000 atmospheres) and, because of the mass circulation of water along deep-sea currents, there's no shortage of oxygen. It's also been around a very long time and there are few barriers to dispersal.

In the vast darkness it's not easy to find food or mates. To help overcome these problems chemical signals play an important role, and several deep-sea fish have bioluminescent organs that contain symbiotic light-producing bacteria. Deep-sea anglerfish have gone a stage further; once the (smaller) male finds a female he attaches himself permanently to her, even sharing her blood supply. Another consequence of the darkness is the lack of photosynthetic organisms; the communities are therefore dependent on nutrients and energy from dead organisms sinking to the sea floor. These range in size from whales to microscopic plankton. The smaller particles often flocculate, combining with mucus, nutrients, bacteria and protozoans, to form flakes of 'marine snow'. Most of the organic matter is eaten, or a lot of the nitrogen removed, on its journey to the ocean floor, so by the time it gets there it isn't very nutritious. This is one reason why the biomass on the sea floor is very low.

One important area for future research is the role of bacteria in the nutrient dynamics of the abyssal zone.

see also...
Oceans

1

Balance of nature

The 'balance of nature' has been a background assumption in ecological thinking since antiquity. The phrase is rather fuzzy (balance of *what* exactly?), but it generally implies that nature, when left undisturbed, is ordered and harmonious. Throughout history this harmony has been associated with divine providence. For example, according to the Greek philosopher, Herodotus, the reason why predators don't eat all their prey is that divine providence has provided them with different reproductive capacities: timid prey produce abundant young, while 'savage and noxious creatures are made very unfruitful'. In the case of lions, so Herodotus argued, the female can only produce one cub in her lifetime, because as the cub grows inside the womb it tears it to shreds with its claws. Mercifully for lions this isn't true, and it's a rather poor example of divine wisdom, as the species would soon go extinct if two adults produced just one offspring between them.

Occasionally divine retribution upsets the balance, but afterwards nature returns to a state of harmony. A thornier problem for the benevolent creator idea came with the discovery of fossils, and the realisation that they belonged to extinct species. If nature was in harmonious balance, why did species become extinct?

Twentieth century views of the balance of nature range from Charles Elton's opinion that, 'The balance of nature does not exist', to the view that species help maintain the balance for the 'common good'. This latter belief has led to the idea that communities, and even Earth, are 'superorganisms'. No serious scientist would entertain the common good or superorganismic view today, and there has been a definite shift towards Elton's view.

Arguments over the balance of nature have important implications for the way we manage and exploit nature, for example in the way we resist or embrace natural disturbances.

see also...

Disturbance; Equilibrium; Population regulation; Scale in ecology

Behavioural ecology

Why do some animals behave altruistically? And what do we mean by behaviour?

Key assumptions underlying behavioural ecology are that there is a genetic component to behaviour, on which natural selection can act; and that adaptations arise because they are beneficial to an individual *not* to the species. Against this backdrop, behavioural ecologists try to understand the consequences of behaviour for an individual's chances of survival and reproduction, both of which contribute to its fitness. Fundamentally, fitness is about individuals getting copies of their genes into the next generation. Direct reproduction isn't the only way of achieving this; as close relatives share a proportion of their genes, then helping them to reproduce can be a successful strategy under some circumstances (called indirect fitness).

Theory has played a major role in the development of behavioural ecology, particularly models borrowed from economics involving cost-benefit analyses. Of course, an individual's best 'strategy' often depends on what competing individuals of the same species are doing. In addition, although it's comparatively easy to get some idea of the relative costs and benefits within a particular activity, a major challenge for behavioural ecologists will be to determine the trade-offs between different activities, such as foraging for food, finding a mate and avoiding predation.

Recently a major focus of behavioural ecology, due in no small part to developments in molecular biology, has been on reproductive strategies, particularly sexual selection involving mate choice and sperm competition. Behavioural ecology is also increasingly being linked to higher levels of biological organisation. For example, unlike traditional models in population ecology, recent developments in 'individual-based' models of population dynamics explore the impacts that individual differences in behaviour have on populations.

Finally, behavioural ecology has recently begun to be applied to conservation studies; for example, how will a species respond to a change in its habitat?

Biocontrol of pests

Chemical pesticides have had major environmental impacts. Today's pesticides break down quickly in the environment, but they are still designed to kill and a lot of people are concerned about their effects on human health and the environment.

Biological control is often held up as an effective, environmentally friendly and cheap alternative to chemical pesticides, and it has been an important weapon in the battle against insect pests for over 100 years. About 40% of biocontrol agents that become established have been successful.

Insects often become pests because they've escaped control by their natural enemies. Most crops are grown in areas to which they are not native. The insects that feed on the plants back home are often introduced with the crop. It must seem like insect heaven: a huge amount of top-quality food, and not a natural enemy in sight. No wonder they become pests.

A crude caricature of biological control is that you visit the original home of the pest, find out what eats it, test their effectiveness and then, all being well, release them to control the pest; and there have been some impressive successes using this approach; and some disastrous unintended consequences. Some biocontrol agents have become pests themselves, like the cane toad introduced into Australia; others have even caused the extinction of other species. Before introducing non-native species to control pests, the risk to non-target species often needs to be evaluated.

Nowadays, no one would be stupid enough to introduce non-native, generalist vertebrate predators to control a pest. However, there is currently a lot of debate about the risks associated with biological control, particularly against insect pests.

see also...

Parasitoids; Population regulation; Predation; Species introductions

Biodiversity

What is biodiversity? Why is it important? And why should we conserve it? Biodiversity in its most abstract form is simply the 'variety of life'. This encompasses genetic diversity within species, the diversity of species and higher taxonomic levels (families, classes, phyla etc.), and the diversity of habitats and ecosystems. Because biodiversity is expressed at so many different levels there is no single all-encompassing measure of it, hence its rather abstract nature. You can, however, measure bits of it, such as species diversity, and in practice this is often the focus of attention.

Biodiversity has come to mean much more than the variety of life. Not only has it become a field of study, it's also become heavily value laden: biodiversity is good, loss of biodiversity is bad, and we should do all we can to maintain it. This has meant a shift in conservation priorities away from the focus on a limited number of single (usually 'charismatic') species towards a more balanced approach that includes biodiversity considerations. Many reasons have been given for conserving biodiversity. These range

from the belief that the variety of life has intrinsic value and we have moral and ethical responsibilities towards it, to anthropocentric pragmatism: biodiversity provides various 'ecosystem services' (see Ecosystem, p. 29) and there are plenty of economically useful things to exploit from it, from anti-cancer drugs to ecotourism.

How do we go about conserving biodiversity? One approach is to focus efforts on the best examples of the widest range of ecosystems. Another approach is to focus on 'biodiversity hotspots', geographical areas with any combination of the following: lots of species, high levels of rarity or endemism (species found nowhere else) and facing a severe threat. By protecting hotspots you preserve more species than in other areas of similar size.

see also...

Conservation; Ecological redundancy; Ecosystem; Latitudinal diversity gradient

Biogeochemical cycling

Biogeochemistry focuses on the global distribution and transport of biologically important elements, such as carbon, nitrogen, phosphorus and sulphur, and various essential trace metals.

Unlike energy, which arrives from the sun and is eventually dissipated back into space, the elements form a closed system – a cycle – with the atoms being used over and over again. As a lot of the chemicals are soluble in water, their cycling is closely linked to the water cycle, which in turn is driven by the sun's energy. Water flows via rivers to the oceans, where each molecule spends an average of 3500 years before the sun's heat causes it to evaporate (along with 16 million tonnes of other water molecules each second). It then spends a few days in the atmosphere before returning to the surface as rain or snow.

All the 'bioelements' pass through various 'reservoirs', though to differing extents. These reservoirs include the atmosphere, fresh and salt water, soil and rock and, of course, the bodies of organisms. For example, nitrogen has a huge atmospheric component (making up 79% of the air), and some 'nitrogen-fixing' bacteria can use atmospheric nitrogen directly and convert it into a form usable by plants (nitrates). In contrast, phosphorus has no atmospheric component to speak of. It tends to head slowly but inexorably for the sea, where it stays for a few million years, being recycled by countless organisms. Eventually it becomes incorporated into ocean sediments where it stays for a few hundred million years before the ocean floor becomes a mountain and weathering releases it into the biosphere once more.

Every element cycle has its own peculiarities and some, like carbon and nitrogen, are extremely complex. This makes detailed understanding of these cycles difficult, but it is very important that we do understand them because they are being altered by human activity.

see also...

Decomposition; Gaia; Microbial loop

Biomes

The biome concept is an attempt to classify the world's major terrestrial vegetation types, based on the influence of the world's climate.

The two main climatic factors influencing vegetation are temperature and precipitation (rain, snow etc.). There is a clear temperature gradient from the equator to the poles, but there are also modifying factors such as distance from the oceans, the direction of ocean currents (e.g. the Gulf Stream brings warm water to the north east Atlantic) and the presence of high mountain ranges.

Combine high average temperatures found near the equator with high rainfall and you get tropical rainforests. At the opposite end of the spectrum, low precipitation and low temperatures lead to tundra. Warm, dry regions contain deserts. A bit more rainfall and you get savannahs, and so on.

At this global scale the dominant 'life forms' of the vegetation, rather than species, are recognised (this distinguishes biomes from communities). To take just one example, succulent thorny plants are widespread in deserts, but they are made up of cacti in the New World and completely different families in the Old.

Within biomes, differences in topography, geology, soil moisture, disturbance etc. determine the local communities found in different areas; indeed in some cases they may override the effects of the climate. Then there's the effect of altitude; even in the tropics the highest mountains are treeless and snowcapped.

What about aquatic communities? These are less affected by climate, particularly the oceans, which have far smaller temperature fluctuations than those experienced on land. The dominant life forms depend more on local conditions (e.g. depth and flow of water) than position on the planet.

see also...

Coniferous forests (taiga); Deserts; Grasslands; Life forms; Mediterranean shrubland; Savannah; Temperate forests; Tropical rainforests; Tundra

Chaos

Chaos has been described as one of the 20th century's most important scientific advances. Its world of 'strange attractors' and 'fractional dimensions' has fundamentally altered the way in which scientists view the universe.

If a system were purely deterministic, with no random elements, then you'd expect that it would be easy to make accurate predictions about the system. Not necessarily. It's been shown that very simple, purely deterministic processes can result in exceedingly complex, seemingly random, fluctuations – deterministic chaos.

The hallmark of chaos – deterministic or otherwise – is extreme 'sensitivity to initial conditions'; in other words the most minuscule differences at the start are amplified at an ever-faster rate, leading to wildly different trajectories. This means that although chaotic systems are predictable in the short term (like the weather) they become increasingly unpredictable over longer time spans (this contrasts with purely random systems, which are equally unpredictable over any time scale).

In the case of a simple model of density-dependent population growth, as the number of surviving offspring produced by each individual increases, the behaviour of the population changes from a stable equilibrium, to periodic cycles, to, under very high growth rates, chaos. Even here, the population size is bounded (stable) – it doesn't keep growing indefinitely or go extinct – but within these boundaries the behaviour of the population is largely unpredictable.

Chaotic dynamics have been found in simplified, single-species laboratory systems, but what role is there for deterministic chaos in the highly variable, 'noisy' natural world? Unequivocal evidence of chaos in the field is lacking. This doesn't mean it's not there or that it's unimportant, just that under field conditions deterministic chaos isn't easy to distinguish from, or is drowned out by, population fluctuations arising from density-independent environmental variation (noise).

see also...

Density dependence; Equilibrium; Metapopulation; Population regulation

Character displacement

When two species compete for a share of a limiting resource it's easy to imagine that there would be strong selection pressure to lessen or avoid the effects of competition. One way of doing this is by actually evolving an increase in competitive ability and 'competitively excluding' the other species; the other way is evolving to minimise the overlap in resource use (and hence competition). The latter process is known as ecological character displacement.

Direct experimental evidence of character displacement isn't easy to come by, so most of the evidence is circumstantial, based on patterns observed in nature. Unfortunately, several different processes can cause the same pattern, and distinguishing between them can be very difficult. This is a perennial problem in ecology.

One pattern thought to result from character displacement is when two closely related (and presumably competing) species are morphologically more different when living in the same area than they are when living alone. Such patterns

certainly exist, but it's far from easy to show unequivocally that they result from the evolutionary process of character displacement. In order to demonstrate that character displacement has occurred several conditions need to be met. Among other things, you'd need to show that the trait is inheritable, that the species are in fact competing, and that the character you're looking at reflects differences in the use of the limiting resource. Few, if any, studies meet all the necessary criteria.

Another pattern sometimes seen among closely related co-occurring species, are larger than expected differences between body sizes, particularly bits of bodies relevant to food acquisition (like teeth). Not unreasonably this pattern has been interpreted as evidence of interspecific competition, although this is still a matter of debate.

see also...

Coevolution; Guilds; Interspecific competition; Semantics; Species coexistence

Chemical ecology

The possession of smelly feet is not normally considered a threat to health, unless you live in malaria-infested countries. The mosquitoes that transmit malaria to humans are attracted to the smell of various chemicals emanating from our breath and skin; and people with sweaty feet are particularly attractive (the smell is caused by bacteria that live in our sweat glands).

Chemicals are therefore an important means of deliberate and accidental communication, both within and between species, and can benefit either the sender or the receiver of the signal. And where there is communication there is often deceit. For example, some spiders lure male moths by mimicking the odours emitted by females and some plants trick carrion- or dung-feeding insects into pollinating them by emitting chemicals such as cadaverine, putrescine and skatole.

Chemicals are also important in defence (and attack). Examples include Bombardier beetles that squirt boiling, toxic chemicals at their attacker, and hedgehogs that cover their spines with toxic chemicals obtained from the skin of toads.

As might be expected, however, it is in plants that chemical defences are best developed. Plants possess an incredible diversity of chemicals that apparently are used solely as a means of defence. The chemicals may be highly toxic, act as deterrents, or merely slow down the digestion of herbivores. Nicotine, caffeine and tannins are familiar examples of plant defence chemicals.

Despite their ubiquity, the effectiveness of chemical defences is not always clear. Most plants have a variety of generalist and specialist herbivores to contend with, some of which evolve counter-defences. One herbivore's poison is another's 'meat'. This makes the study of plant chemical defences difficult, and a number of questions remain unanswered.

see also...
Herbivory

Coevolution

In evolutionary terms, species can't afford to stand still. Even if the abiotic environment isn't changing, a species is still subjected to selection pressures from predators, parasites, prey and competitors, which are themselves evolving. These evolutionary changes can be rapid, occurring over what are usually thought of as ecological time scales.

This 'reciprocal evolutionary change in interacting species' (Thompson, 1982) is called coevolution. Here an evolutionary change in one species leads to an evolutionary response in another, which in turn leads to a further response by the first species, and so on. Examples of coevolution include some insect pollinators and plants (fig wasps and yucca moths being classic examples), mimicry, symbiotic mutualists, and the evolutionary 'arms race' between parasites and hosts, and predators and prey.

For example, when the *Myxoma* virus was introduced into Australia in the 1950s to control rabbits it killed over 99% of the population. This intense selection pressure led not only to the rapid evolution of increased resistance of the rabbits to the virus but also to decreased virulence of the virus.

Many classic examples of coevolution are highly specialised symbiotic (mutualistic and parasitic) relationships between species; although we must be careful not to assume that coevolution is the process leading to the specialisation. With less intimate associations, evidence for reciprocal coevolutionary changes is more ambiguous, largely because it's not always clear to what extent herbivores affect the fitness of plants.

see also...

Character displacement; Mutualism; Parasitism; Symbiosis

Community

A community is simply a collection of species occupying a particular place. Sometimes the particular place appears to have reasonably well-defined boundaries, e.g. a pond, sometimes not, e.g. in grasslands species gradually replace each other along a moisture gradient. Of course, any boundaries are more a reflection of human perception than ecological reality, for example, the boundary between a pond and the surrounding land might seem real to us, but it's not to many of the plants, amphibians, mammals and aquatic insects that live there. Similarly, a community can be any size, from the community of organisms inside a termite's gut, to the organisms living in the vast African savannahs. Ecologists rarely study the entire community – there are simply too many species – so for pragmatic reasons some subset of species is the focus of study, such as guilds or taxonomic groupings, e.g. spiders.

In many ways, then, the community as a physical, bounded entity is an artificial construct. Communities are therefore better thought of as a level of biological organisation. Ironically, one thing not made explicit in the definition of community given here is the thing that makes community ecology interesting, the fact that species within a community interact (often indirectly) with each other. The frequency and strength of interactions between species produces a continuum of community types. At one extreme the community consists of recurrent associations of coevolved, strongly interacting sets of species; at the other extreme we have looser associations of species that simply live in the same habitat because they have similar environmental requirements (known as the individualistic concept).

Historical ecology shows us that membership of communities is rather flexible, with species coming and going as environmental conditions change, and the prevailing view among ecologists is that communities lie more towards the individualistic end of the spectrum of community types.

---

see also...

Community assembly; Community structure; Generality in ecology; Guilds; Species interactions

Community – alternative stable states

In recent years algal blooms have become a regular occurrence in lakes. These blooms are caused by increased levels of plant nutrients particularly nitrates and phosphates – a process known as eutrophication. Eutrophication can lead to a species-poor ecosystem, where a lake dominated by algal blooms increases the demand on the available oxygen, which in turn leads to the death of other species.

Originally it was thought that preventing any more phosphorus from entering a lake would restore it. However, this often failed in shallow lakes. In such circumstances communities are capable of surviving in two 'alternative stable states'. Healthy plant populations dominate the 'normal' state. Even in the presence of increased phosphate levels, algae are prevented from taking over by introducing several 'buffer' mechanisms, such as grazing by zooplankton (in this case, tiny crustaceans); hence this state is 'stable' in the sense that it is resistant to change.

With all this buffering going on, why do eutrophic lakes still end up in an algae-dominated state? The answer is that a shock to the system – such as pesticides killing the zooplankton, or the destruction of plants – can flip the lake from one state to another. This second, algae-dominated, state is also self-sustaining. The algae shade out the plants, and therefore there is nowhere for the zooplankton to hide from predators, leaving nothing to reduce the algae populations. In this case reducing phosphorus inputs has no effect. The essence of the alternative stable state idea is such that even if the factor that had caused the shift to the alternative state is lessened or removed, the system doesn't return to its previous state – it has crossed a boundary.

Another example comes from the Serengeti–Mara ecosystem. Here fire can flip the ecosystem from woodland to grassland, which is maintained by grazing and browsing. It may flip back to woodland following outbreaks of disease among the grazers, which in turn allows tree seedlings to grow.

see also...
Community; Trophic cascade(s)

Community assembly

How are communities assembled from the pool of available species in a geographical region? Are they merely random subsets of species from the regional pool, or are there 'rules' that 'forbid' certain combinations of species? Are species-rich communities better able to resist invasion by new species? What effect does the order of arrival of species have?

Many ecologists view the regional pool simply as a source of potential colonists, arguing that, as it is created by long-term evolutionary and geological processes, e.g. extinction, speciation and continental drift, they aren't directly relevant to the assembly of communities.

The actual species composition of a community is the result of a number of processes (constraints) operating at progressively smaller spatial and temporal scales. These act like a series of filters that prevent entry by some species while allowing others through. The first filter is the ability of species to arrive, so distance from the regional pool, time, and dispersal ability play a role here. Having arrived, the species has to cope with the environmental conditions and also be able to coexist with other species in the community.

A difficulty in trying to study the assembly of communities is that, to varying extents, they are the products of past events. An alternative approach is to study aspects of community assembly in the laboratory. This makes it possible, for example, to manipulate the arrival sequence of species to see if it makes a difference to community structure (apparently it does).

Humans are currently changing the nature of the filters, we are adding and subtracting species to regional pools and communities, climate change will modify regional environmental conditions, and habitat fragmentation will affect dispersal. Communities are going to disassemble and reassemble with increasing frequency and with unknown consequences.

> ### see also...
> *Community; Scale in ecology*

Community structure

What processes determine the number, identity and relative abundance of species within a community? What processes determine community structure? To what extent is community structure determined by local processes operating within the community (interactions between species, disturbances) compared with processes operating at the regional level (habitat selection, dispersal ability)?

Membership of a community is clearly limited, not all species in a region are found in a single community. Community membership is determined by three factors that operate in a hierarchical fashion: there's the ability to get there, the ability to survive under the prevailing environmental conditions, and the ability to coexist with other species.

Until recently the focus was on within-community processes. A vociferous debate between community ecologists in the 1970s and 1980s centred on the relative importance of interspecific competition, disturbance and predation as the dominant forces

structuring communities. It was argued, for example, that interspecific competition was widespread and was the dominant process structuring the community. Other ecologists asserted that predation and disturbance kept the populations of many species below a size at which competition would be important. Both extremes contained an element of truth; competition, predation and disturbance are all important, but to varying and largely unpredictable degrees in different communities.

More recently the approach to understanding community structure has been to take a step back from the nitty-gritty of interactions between species, to look at how the number of species (species richness) in a community is affected by regional processes. Although more evidence is needed, it seems that the number of species in some communities is directly proportional to the number available in the regional pool.

see also...

Community; Community assembly; Succession

Complexity and stability

Until the early 1970s it was largely assumed that the greater the complexity of a community, the greater its stability (see Equilibrium, p. 33 for a brief discussion of stability). Here complexity, roughly speaking, is a combination of the number of species, and the frequency and intensity of interactions between them. The positive link between complexity and stability makes intuitive sense – natural systems do seem to be quite complex – and this intuition could be justified by arguing that the more species there are in a community and the more links there are between them, will inevitably lend more help in 'soaking up' the effects of any perturbation(s) to the system.

Then, a series of influential theoretical models indicates that the opposite was true, i.e. that more complex communities are less stable. However, models are only as good as the assumptions on which they are based and a number of assumptions relating to these models were rather unrealistic. Recently, more realistic models have focused on the importance of 'weak' interactions between species. It seems that it makes a big difference to community stability, depending on whether interactions between species are consistently weak, which has a stabilising effect, or whether the interactions are weak but highly variable, which will tend to magnify any perturbations. So much for theory, what about experimental evidence?

What are we to make of all the conflicting theories and experimental evidence? The link between complexity and stability depends, among other things, on how stability is measured, the nature of the perturbation, the strengths of, and variations in, the interactions between species, and the environmental setting. May, one of the more influential figures in the debate, has recently said, 'As the programme of research continues to unfold, it defies any easy generalisation' (May, 1999). A statement that could be applied to most, if not all, aspects of community ecology.

see also...

Ecological redundancy; Equilibrium; Food web; Models in ecology

Coniferous forests (taiga)

The taiga is on the move. These vast, biomonotonous northern coniferous forests, dominated by pine and spruce, have been moving northward since the ice-caps retreated after the last glaciation.

The region is characterised by bitterly cold winters (down to −40°C) and relatively mild summers (reaching 10–15°C). And it snows a lot. The Christmas-tree shape of the trees helps them shed snow, as do the needle-leaves, whose shape also helps to reduce water loss. (Water availability is a problem as it's locked up as snow or in the permafrost for most of the year.) And, by remaining green all year, the trees are ready for photosynthetic action whenever the temperature allows.

Because of shading, there are relatively few understorey plants. There is, however, a dense carpet of needles decomposing extremely slowly in the cold conditions. Over the millennia these forests have accumulated vast stores of carbon in the trees, litter and soil.

The build-up of fallen needles makes these forests prone to fires. Some species are well adapted to these fire-prone environments. Indeed, some have evolved a degree of fire-dependency; for example, some pines require heat from a fire to melt the resin sealing the cones, thereby releasing the seeds. In the Canadian coniferous forests jack pine and aspen (which can sprout from its roots) are the first to grow back after fire. These dominate the forests for a few years, until spruce overtakes them.

Over the last century the average temperature of northern coniferous regions has risen by about 2°C: as it gets warmer, the forests get drier, leading to more fires and the release of CO_2 (a key 'greenhouse gas') into the atmosphere. If the fires are hot enough to burn deeply into the soil, they will release the vast amounts of carbon stored there.

see also...

Fire; Global environmental change; Tundra

Conservation

In 1979 the Large Blue butterfly became extinct in England following decades of decline. Detailed research into its ecology determined the reason for its decline, although it came too late to save the British populations. It was found that the caterpillar is dependent on a single species of red ant, which it parasitises. In England these ants only survive on warm, south-facing slopes with very short turf. A combination of factors, including changes in farming practices, led to a decline in grazing. The ants subsequently declined in abundance, and were quickly followed by the butterfly. Armed with this understanding, the Large Blue has since been re-introduced, and is doing well.

This example illustrates an important point: in conservation there is no substitute for detailed ecological understanding – assuming there is time to obtain it. Conservation biology has been called a 'crisis science' in which crucial decisions have to be made based on incomplete knowledge. In addition, various other dimensions of conservation can't be ignored.

The aim of biological conservation is to, 'ensure the continuing existence of species, habitats and biological communities and the interactions between species and with ecosystems' (Spellerberg, 1996). Conservation involves a variety of approaches. In the past the focus was often on particular (often charismatic) species, although the broader goals of conserving habitats and functioning ecosystems have achieved greater prominence in recent years. A mainstay of conservation is the setting aside and management of habitats.

A last-ditch conservation technique is captive breeding of species in zoos, followed by re-introduction back into the wild. Again, success depends on detailed ecological knowledge, and ensuring that the reason for the species' decline no longer operates in the area of release.

see also...

Biodiversity; Habitat management; Habitat (re)creation; Habitat restoration; Minimum viable population; Rarity

Decomposition

In most ecosystems the majority of plant biomass never actually gets eaten by herbivores, instead passing straight to the decomposer food chain. Terrestrial ecosystems contain almost twice as much dead plant material (detritus) as plant biomass, with vast hidden armies of organisms consuming it. However, perhaps because most of the process is hidden below ground or under water, decomposition hasn't been given the attention it deserves, even though some communities are sustained by the input of dead organic matter. For example, the main food source sustaining life on the ocean floor and many forested streams comes from dead things, ranging from leaves falling off riverbank trees, to deceased whales and plankton sinking to the sea floor.

Decomposition occurs by both physical and biological means. Decomposers (prokaryotes and fungi) break down complex organic material into simpler inorganic forms that are then available for plant uptake, a process called mineralisation. This is obviously crucial in the recycling of nutrients. Although detritivores feed directly on decomposing plant material and/or the microbial decomposers living on the material, they don't mineralise the detritus. However, their feeding activities reduce the size of the fragments, increasing the surface area for microbial attack. Decomposition is therefore most efficient when both decomposers and detritivores are present.

Dead plant remains are not very nutritious: wood is full of hard-to-digest structural substances such as cellulose and lignin, and deciduous trees withdraw most of the nutrients from leaves before they fall. Why so few animals have evolved the ability to digest cellulose remains a mystery, as it's a major component of plants; instead they rely on bacteria and protozoans to digest it for them. In contrast, dead animals are very nutritious so they're in demand from a range of carrion eaters, and are often eaten before they have a chance to decompose.

see also...

Ecosystem; Microbial loop; Primary production

Density dependence

Studies of American prisons in the 1970s showed that the more crowded the inmates, the higher the frequency of prisoner misconduct and the higher the death rate. Both misconduct and death rate are therefore dependent on the population density, i.e. the number of individuals in a given area, volume, or unit of habitat such as a leaf. There are many examples of animals and plants which have a higher death rate in more crowded populations, with the brunt of the mortality often falling on the young. For example, when there were fewer than 80 female deer on the island of Rhum the mortality rate of juvenile males was almost zero; with 160 adult females, the death rate shot up to over 60%.

A similar (but inverse) pattern is seen with birth rates. The more birds there are in a given area, the *fewer* eggs, on average, are laid by each female. The same idea applies to immigration and emigration. So, as population size increases, density-dependent factors result in the death or emigration rate increasing, and/or the birth or immigration rate decreasing. Density-dependent factors are 'biotic' (caused by living organisms), and include predation, parasitism, and competition within and between species.

The impact of some factors affecting birth and death rates bears no relation to how crowded the population is. For example, a bout of frost might kill 10 out of a population of 100 insects, and 50 out of a population of 500. Although the absolute *number* of deaths will be bigger in the larger population, the *proportion* dying is the same at both densities. Frost and other abiotic (non-living) factors are therefore density-independent factors.

The importance of density-dependent factors lies in the fact that it is only these that can keep populations within certain bounds (i.e. regulate the size of a population), by acting as a brake on population growth rate as the population density increases.

see also...

Equilibrium; Intraspecific competition; Population; Population regulation

Deserts

Deserts generally lie around the Tropics of Cancer and Capricorn, and occupy about a third of the Earth's land surface. Within these regions, factors tending to lead to the formation of deserts include lying in the 'rainshadow' of mountain ranges like the Rockies and Andes; or being a long way from the ocean (the main source of moist air).

When we think of deserts we usually think of hot, arid places, but some deserts can get quite cold, even dropping to around freezing at night; the Gobi is one example. Even hot deserts get quite cold (though rarely freezing) at night due to the lack of cloud cover. Recurrent features of deserts are the paucity of rain – less than 25 cm a year – and the high evaporation rates.

Plants cope with the infrequent, unpredictable rain by one of the two basic methods: avoid or tolerate the arid conditions. Some plants appear from nowhere (actually, from dormant seeds or underground bulbs) once the rains come, and complete their life cycle or replenish their underground stores in just a few weeks. The second strategy involves being very efficient at obtaining, and retaining, water. A lot of plants have extensive, shallow root systems to maximise water uptake when it rains, which leads to a rather regular spacing of plants. Succulent plants, like cacti, store water taken up during the rains. Anything that stores water in an arid environment is likely to attract the attention of animals; hence the widespread use of spines and chemical defences shown by many desert plants.

As the vegetation is both sparse and largely indigestible, the species diversity of herbivores is low. In contrast, seed-eating animals and their predators are common, as are lizards, which are well adapted to living in warm, dry conditions.

Whereas a lot of the other biomes are decreasing in size due to human impacts, deserts are on the increase, particularly in Africa where the rapidly expanding population, overgrazing, and the removal of what little wood there is, is leading to desertification.

see also...

Mediterranean shrubland

Disease

The 1990s brought us heroin-chic; in the 19th century we had TB-chic. Many romantic heroines in novels died of consumption, and the consumptive look became quite fashionable. Today, two billion people worldwide are infected with the tuberculosis bacterium, and it is the commonest fatal infection in humans. TB has been increasing over recent years, even in developed countries, due to a combination of evolved antibiotic resistance and an increase in poverty and crowding in cities.

Of course, TB isn't the only disease to have huge effects on human populations: up to 50 million people died in the flu pandemic of 1918–19, many more than were killed during World War I. And then there's bubonic plague, malaria, smallpox.

Although most human diseases have been around for some time, several have become more prevalent since humans started living in large, crowded cities. Measles, for example, needs a population of about 300,000 before it becomes an epidemic. These 'transmission thresholds' are important when designing vaccination programmes: you immunise just a high enough proportion to keep the number of susceptible individuals below the threshold.

Diseases have major impacts in natural communities, particularly when a bacterium or virus jumps between species, as the new host has little or no immunity. This jump often happens when humans and their domestic animals colonise new areas. For example, the rinderpest virus, introduced into the Serengeti by domestic cattle, killed over 80% of some ungulate (buffalo, wildebeest etc.) species which in turn led to a reduction in the populations of their predators. Following vaccination of the cattle, the numbers of ungulates and their predators have increased.

Diseases also have implications for conservation. For example, canine distemper, caught from stray dogs, nearly wiped out the few remaining black footed ferrets in North America.

see also...

Parasitism; Savannah

Dispersal

At the end of the 19th century European starlings were introduced into Central Park in New York, by the end of the 20th century they had spread across the entire country. An impressive example of dispersal.

Dispersal – the spreading of individuals away from each other – is an important factor determining the distribution of organisms, and the ability to disperse is almost ubiquitous – exceptions include flightless birds. Even sedentary organisms like plants and barnacles disperse, as do seeds and planktonic larvae. It's easy to see why organisms disperse in unstable habitats, or in habitats undergoing succession, but there are good reasons for dispersal even in stable habitats. It's a way of reducing the chances of inbreeding and can provide an escape for individuals from competition with their kin.

Dispersal ability can also play a part in allowing competing species, or a predator and its prey, to coexist. There is often a trade-off between competitive and dispersal ability so, as long as new habitats are created frequently enough, the weaker competitor may survive by having a fugitive lifestyle in which it keeps one step ahead of the arrival of the stronger competitor. Dispersal is the key process in the ecology of metapopulation, and is particularly important in the context of a species' response to habitat fragmentation.

Dispersal can be active (flight) or passive (wind-blown seeds). It's often in the young stages that dispersal occurs, although in terrestrial insects the adults also disperse. Dispersal ability entails costs, particularly active dispersal, and it also carries risks, i.e. the dispersing organism might not find a suitable habitat.

Dispersal ability places an important limit on the distribution of many species. However, by moving thousands of species around the world, humans have removed this limit, with serious ecological consequences.

see also...

*Life history strategies;
Metapopulation; Patchiness;
Species introductions*

Disturbance

When the great storm of 1987 hit England it uprooted 15 million trees and was considered a national disaster. Money donated to plant new trees was largely unnecessary, however, as woodlands regenerate naturally.

Disturbances occur over a large range of frequencies and magnitudes, with these two aspects usually being inversely related. At one end of the spectrum we have frequent, small magnitude disturbances, such as children turning over rocks on rocky shores; at the other end there is the very occasional, but rather intense, effects of a collision with a meteorite.

It's not easy to define disturbance in a meaningful, non-arbitrary way. What we call disturbance tends to be a relatively discrete (as seen by us) event that isn't targeted at particular organisms, which leads to some mortality, releasing resources such as space as a consequence. Because disturbance is inevitable in most habitats, species have evolved strategies to cope with it, with many actually relying on disturbances to free up space.

Disturbance is an important means by which species diversity is maintained. Succession occurs in the gaps created by disturbance and so, at a regional level, habitats consist of a shifting mosaic of vegetation patches at various stages of recovery, each with its particular blend of species.

At a more local level there is the 'Intermediate Disturbance Hypothesis'. This states that when disturbances are small and infrequent the competitively dominant species will eliminate the weaker competitors, hence diversity will be reduced. If disturbances are large and frequent, however, they will tend to decimate the community, again leading to a reduction in diversity. Thus maximum diversity is predicted at intermediate levels of disturbance.

see also...

Balance of nature; Fire; Mediterranean shrubland; Succession

Ecological energetics

The fundamental currency of the universe is energy, with the ultimate source of energy on Earth coming from the sun. Not only is its light energy the driving force behind ecosystem processes, but its heat drives the water cycle (among other things) which in turn is intimately linked with the cycling of nutrients. Ecological energetics is concerned with the flow, transformation and efficiency of use of energy through organisms and ecosystems.

Of all the energy in sunlight that reaches the Earth's surface less than 1% is converted into net primary production by plants. There are several reasons for this: for example, a lot of the light simply bounces off the plants, and the energy from the majority of sunlight wavelengths isn't usable by plants. Once energy is 'fixed' by plants it isn't only used to accumulate biomass: there are the general metabolic costs of staying alive and reproducing.

On average, only about 10–20% of terrestrial net primary production is eaten by herbivores, though the figure is much higher (about 80%) in aquatic ecosystems. This is partly because most terrestrial plants contain a lot of worthless ingredients, ranging from indigestible compounds used as structural support, to toxic anti-herbivore chemicals.

In contrast, there is little of their vertebrate prey that carnivores don't consume. With insect prey, because of their small size there's a higher proportion of indigestible bits like the exoskeleton. The cost of maintaining a constant body temperature is a particularly important factor in birds and mammals, resulting in only 1 or 2% of assimilated energy ending up as new biomass.

Taking all these losses and inefficiencies into account, between 2 and 24% (with an average of 10%) of the energy entering one trophic level is transferred to the next.

see also...

Decomposition; Ecosystem; Microbial loop; Primary production; Trophic level(s)

Ecological redundancy

It is estimated that the current rate of species extinction is between 100 and 1000 times the rate that was prevalent before humans appeared. An important question is how the loss of species affects the functioning of ecosystems. By functioning we mean primary production, decomposition, nutrient and water recycling etc.

The link between diversity and ecosystem functioning is an area of intensive research and debate at the moment. At either end of a continuum we have the rivet and the redundancy hypotheses, with the idiosyncratic hypothesis somewhere in between. The rivet hypothesis likens species to the rivets holding a plane together in that each rivet has a small but important role, and the plane is progressively weakened as rivets are removed. Eventually a point is reached when the plane (ecosystem) falls apart. The redundancy hypothesis states that most ecosystem functions are carried out by several species thus, beyond a minimum diversity, the loss of a few species has little or no effect. The idiosyncratic hypothesis states that the link between diversity and ecosystem functioning is largely unpredictable.

So which hypothesis is right? Probably all of them, depending on the ecosystem examined, the ecosystem processes studied, and the temporal and spatial scale of the investigation.

A worry some people have is that labelling species as redundant might imply that it doesn't matter if they go extinct. We would never have enough information to say whether a species was functionally redundant in all possible ecosystem processes, and who's to say what role it might play when an ecosystem is faced with occasional large disturbances? Most other environmental impacts caused by humans are, theoretically at least, reversible. The extinction of species is unique once a species is gone it is gone for good, so there is a clear need to apply the precautionary principle.

see also...

Complexity and stability;
Functional groups; Keystone species

Ecology

In some ways ecology is as old as humans. People have always depended on knowledge of their environment, and the animals and plants that live there, in order to find food, to hunt and to grow their crops successfully. It was only towards the end of the 19th century, however, that ecology became a 'self-conscious' scientific discipline, in other words ecologists began to recognise that what they were doing was ecology (rather than some obscure branch of physiology), and started to call themselves ecologists.

Charles Elton gave an early, and perhaps the shortest, definition of ecology as 'scientific natural history'. If scientific is taken to mean a scientific approach, then this is a reasonable reflection of what a lot of ecologists actually do. However, while natural history is a crucial element of ecology, it's not synonymous. You might say that knowledge of natural history is a necessary but not sufficient requirement to becoming an ecologist. Ecology, as a scientific pursuit, is about finding general patterns in nature among all the detail, it's about searching for *explanations* for those patterns. Ecologists delight in the detail but try not to be overwhelmed by it.

Pluralism is the key to ecology. As ecology has matured as a science, it has become apparent that there are unlikely to be clear-cut, universal answers to many ecological questions. Ecological patterns are often caused by several factors that interact with each other, and different processes vary in their importance at different scales of space and time. Ecology, unlike other natural sciences, also has an historical component, for example, many temperate communities are still recovering from the last Ice Age. To understand the natural world we therefore need to use a variety of approaches at a variety of scales.

As humans place increasing pressures on the environment, ecological understanding will be vital if we are to have any chance of a sustainable future.

see also...

Experimental ecology; Generality in ecology; Scale in ecology

Ecophysiology

Woolly bear caterpillars live in the Arctic and can survive temperatures of −70°C. They cope with most of their body fluids being frozen for 11 months of the year, and manage to grab a quick bite to eat during the one month in the summer where the temperature rises sufficiently to thaw them out. It's not surprising that these 'freeze tolerators' take 14 years to reach adulthood.

Other animals survive freezing temperatures by using, among other things, antifreeze proteins to 'supercool' their body fluids to prevent ice forming (freeze avoiders). Both strategies are effective under certain environmental conditions, as long as the fluid *within* cells doesn't freeze; no animal can survive that. The details of the mechanisms used by freeze avoiders and tolerators are fascinating, and the same is true when we look at the ways in which animals, plants and micro-organisms cope with high temperatures, lack of oxygen, drought, pH, water balance and toxic chemicals.

Ecophysiology, therefore, is the study of the performance of individual organisms in relation to their abiotic (non-living) environment.

These abiotic factors act as constraints on where a particular species is theoretically able to survive; they define an organism's 'fundamental niche'. Within these environmental constraints, where a species actually lives is largely determined by its ability to arrive there and, once it's arrived, by interactions with other species (its 'realised niche').

Ecophysiology is relevant to a number of applied issues, particularly agriculture, the ecological restoration of derelict land, and studies of pollution. And, with rapid global environmental change currently under way, a knowledge of ecophysiology is taking on renewed urgency as we try to understand and predict the responses of organisms to changes in climate.

see also...

Niche

Ecosystem

An ecosystem is the community of living organisms plus the non-living environment. It's not a higher level of organisation than the community, just a broader perspective. Ecosystems can be thought of as energy-processing systems in which the inputs – energy, nutrients, water, oxygen – are transformed and transferred by organisms. It's the interactions between the living and non-living components that constitute the ecosystem; not only does the environment influence the organisms but organisms affect their non-living environment.

Ecologists often talk about ecosystem functioning, that is primary and secondary production, decomposition rates, nutrient cycling etc. As more species become extinct a question currently being addressed is the effect a reduction in biodiversity will have on ecosystem functioning (see Ecological redundancy, p. 26).

From a human perspective, ecosystems perform a number of 'services': they provide our food, regulate climate, cycle nutrients, clean up a fair bit of our waste, and so on. Of course, at one level their value to us is infinite as we would die without them. However, as we continue to exploit and pollute, developers and policy makers don't give enough (or any) weight to these services. Rather than fighting economic arguments with ethical or philosophical ones, recent developments in ecological economics attempt to give a monetary value to the products and services provided by nature.

One estimate is that ecosystems provide services worth over US \$33 trillion a year, which is about twice the global gross national product. The exact figure is unimportant; the main thing is that it's huge and that it's largely outside the market and so plays no part in economic decision-making processes. The idea of putting a monetary value on nature and its services is crude, revolutionary and potentially dangerous. But whatever the merits or dangers, it is making economists, environmentalists and policy makers talk to each other, which can only be good.

see also...

Community; Ecological energetics; Ecological redundancy; Ecosystem engineers

Ecosystem engineers

Parts of the Negev desert are covered by a black crust of soil and sand glued together by chemicals secreted by colonies of various species of micro-organism. It's thought that the crust made by the colonies protects them from the intense heat.

When it rains, the hardened surface increases water run-off, which collects in small depressions in the sand created by various desert animals. The moist sand in the pits provides an excellent site for seed germination; the result is the formation of mini-oases containing several species of plant.

By altering the environment for their own needs, the micro-organisms inadvertently benefit a host of other species, and are examples of what have become known as ecosystem engineers. Rather than acting directly as a resource themselves, engineers modify, maintain or create habitats, thereby controlling the availability of resources to other organisms. The effects may be large or small, beneficial or detrimental to others.

Examples include physical engineers: earthworms modify soil structure and affect nutrient cycling; beavers create lakes by building dams; trees alter local humidity, light levels and temperature. Other organisms are chemical engineers: ocean plankton release chemicals that lead to the formation of clouds; ancient micro-organisms released oxygen as a by-product of photosynthesis, making the atmosphere what it is today.

The world is full of ecosystem engineers. All ecosystems are to some degree affected by them; they play vital roles in the process of succession; indeed the very existence of some ecosystems depends on them. Ecologists use them when restoring habitats; plants are often the best way of stabilising soils, and added earthworms and nitrogen-fixing plants to increase their fertility. Bringing together numerous examples under the rubric of ecosystem engineers has alerted ecologists to the pervasiveness of organisms altering their environment for their own needs and consequently affecting other species.

see also...

Gaia; Habitat restoration; Succession – primary

Ecotoxicology

Ecotoxicology is the study of the harmful effects of chemicals on ecosystems. When dealing with something as complex as ecosystems, this is far from easy.

One thing to clear up at the outset: everything is toxic at high enough doses. So we shouldn't worry about reports of pesticide residues on fruit any more than we should about the presence of caffeine in coffee; in both cases it's the *amount* that we should be concerned about.

There are two factors determining the impact of toxicants: their intrinsic toxicity, and the chances of organisms being exposed to them. The toxicity aspect is assessed in laboratory-based toxicity tests. When testing chemicals for human safety we use rats as surrogates, the assumption being that we can extrapolate the effects on one mammal species to another, but just to be on the safe side a sizeable safety factor is applied. But to what extent is it valid to extrapolate from, say, a single species of water flea (a standard organism in toxicity testing) to the thousands of other invertebrate species living in lakes and rivers?

There are many other problems with single-species toxicity tests, and to get round some of them we can study the effects of chemicals on communities in the laboratory or field. This certainly adds realism, and these tests are effective if we want to study the fate of chemicals, but they are too expensive and time-consuming to test more than a handful of chemicals. What measure of harm should we use? Should we be looking at aspects of ecosystem structure or function? What aspects? Could we even detect any effects over and above the natural 'background' variation?

Ecotoxicologists must combine a variety of approaches to address all the questions that need answering, for example, chemical and mathematical modelling, single- or multi-species experiments or the use of organisms as 'biomonitors'.

see also...

Biogeochemical cycling; Ecosystems; Experimental ecology

Environment

An organism's environment is made of of four interacting components: physical habitat; other organisms; resources; conditions.

A resource is something that can be depleted, e.g. food, light and space, whereas conditions are physical and chemical features of the environment, e.g. temperature, wind speed, pH and water flow, that can't be used up.

Some environments are described as extreme, for example the freezing Arctic, the burning desert, and the incredible pressures at the bottom of the ocean, although this is determined from a human point of view, rather than from the perception of the organisms that live there. To deep-sea fishes an extreme environment would be the surface waters. The only experience of these they are likely to get is when humans haul them up from the ocean depths, whereupon they might feel a vague sense of unease before exploding due to the pressure changes.

All environments change over a huge range of time scales, from seconds to aeons, and all species have a range of tolerance to this environmental variation that plays a part in determining their distribution and abundance.

The term environment also has several popular meanings, coloured by religious and cultural factors. For some the environment is detached from their everyday experience, others consider themselves to be an integral part of it. Some people make a distinction between the natural environment and man-made environments, others don't.

Perceptions of the environment are important as they determine how we exploit or manage it. They have also changed through time. Early hunter-gatherers probably saw the environment in an adversarial sense, with the advent of agriculture and settlement the environment was perceived as a resource to be exploited. Since the Industrial Revolution and the explosion in population size, the environment has been seen as a convenient dumping ground.

see also...

Global environmental change

Equilibrium

quilibrium means balance. So if the death rate in a population exactly balances the birth rate the population size will remain constant (assuming that the environment remains constant and there's no migration). What happens when there is a disturbance, such as a drought or a large influx of individuals? If the population returns to the equilibrium population size then we have a stable point equilibrium. An equilibrium can be locally or globally stable, for example, a population that returns to equilibrium following a minor disturbance, but not from a larger one, is locally stable but globally unstable. There are several facets of stability: it can involve resistance to change in the first place, or the ability to bounce back quickly after disturbance (resilience).

Environments aren't constant, so a stable point equilibrium for a population is unlikely, and even if the environment were constant chaos theory predicts that under some conditions populations can fluctuate unpredictably without a single equilibrium point. Ecologists therefore now think of a stable

equilibrium as a band of population sizes to which a population returns after a disturbance (known as an attractor).

The debate about whether or not populations and communities are best thought of as equilibrium or non-equilibrium systems largely disappears when it's realised that it depends largely on the spatial and temporal scales studied. For example, a stable metapopulation consists of a number of unstable sub-populations linked by dispersal. A pest population may show a dramatic increase in population size over a relatively short period of time, followed by an equally dramatic crash, but if monitored for long enough it may be found to be fluctuating around an equilibrium level. However, over even longer time periods it's questionable whether a true equilibrium exists since all populations and communities are subject to change, for example, climatic fluctuation.

see also...

Balance of nature; Island biogeography theory; Population regulation; Scale in ecology

Experimental ecology

Experiments play a crucial role in science. They are used to test hypotheses by manipulating the factor(s) you're interested in while keeping all other factors constant (or at least to take into account their effects). Hence experiments are the best method for establishing cause and effect.

Experiments present particular difficulties in ecology, largely because of the huge range of scales over which ecological processes operate, the natural viability of the real world, and also because a multitude of factors interact (often indirectly) to affect a system. We can reflect these complexities by making our experiments more complex, but the more complicated they become the more difficult they are to interpret biologically. It is also important to realise that all experiments involve compromise.

At the precise/unrealistic end of the spectrum we have laboratory experiments, which are good at investigating mechanisms and allow observations over several generations. Lab experiments can be seen as 'quantitative biological models'; a bridge between mathematical models and the complexity of the real world.

Field experiments often involve measuring the response of a system when it is perturbed (e.g. by removing species or adding nutrients), and are a mainstay of current ecology. However; largely because of costs and logistics, a lot of field experiments are restricted in terms of time (2 or 3 years) and space (a few square metres), although this is changing.

It's not always possible to carry out experiments, for ethical or economic reasons, or for reasons of scale. Thus, so-called 'natural experiments', despite their problems, have a role to play too. There is no manipulation, we simply observe what nature has provided (hence they are the most realistic and imprecise) and compare this with the patterns predicted by theory. However, the same pattern can be caused by more than one process, so the link between cause and effect is weak.

see also...

Generality in ecology; Models in ecology

Fire

In the 19th century children and aboriginal Australians were flogged for lighting bushfires. The aim was to suppress fires; the result was an increasing number of more severe fires. Why?

Fire is a natural phenomenon that has played a major role in terrestrial ecosystems for millions of years. Fire occurs frequently under dry conditions when there is sufficient fuel (organic matter), an ignition source, and plenty of oxygen. The two main causes of fire ignition are lightning strikes (over 3 billion strikes each year) and humans.

There are three basic types of fire. Ground fires burn at or below the surface; they can be very destructive with some communities taking centuries to recover. Crown fires burn the tree canopies and can be very intense; they often kill mature trees and, in doing so, create gaps for seedling growth. Surface fires burn fine fuels (pine needles, grass) that quickly turn to ash, producing cool (300°C), fast-moving fires. They kill seedlings and saplings, but mature trees aren't usually severely damaged. Just a few centimetres below the soil surface, temperatures may only be 10–15°C higher than normal, so any seeds or regenerating parts of plants below this depth are protected.

Adaptations of plants in fire-prone environments include fire-resistant bark or underground seed banks. Some plants have evolved a dependency on fire and won't release seeds or germinate unless stimulated by fire. Some have evolved increased flammability. As for animals, most run away, head for water, or burrow.

The Australian aboriginals knew what they were doing. Nowadays the tendency in fire-prone habitats is to leave some natural fires to burn or to use prescribed fire; both reduce the build-up of fuel. There are, however, complex issues surrounding the use of fire in landscapes that have been altered by humans.

see also...

Coniferous forests (taiga);
Disturbance; Grasslands;
Mediterranean shrubland;
Savannah

Food web

If there's one common ecological concept it's the food chain: the transfer of energy through a sequence of organisms in which each eats the one below it in the food chain, i.e. a plant gets eaten by a herbivore, which gets eaten by a carnivore, which in turn gets eaten by a larger carnivore. However, simple linear food chains such as this don't do justice to the complex feeding relationships found in many communities, where most species feed on, and are themselves fed upon by, several species. Feeding relationships between species within a community can be seen as a complex web of interactions – a food web. In diagrammatic representations species are represented as dots, with arrows connecting the dots indicating who eats whom. Although this approach ignores non-feeding interactions, a food web is still a useful way of summarising this important aspect of community structure.

There are three basic types of food web. The first indicates whether feeding links exist between species, it says nothing about the frequency or strength of the links. The second gives an indication of the amount of energy flow between species by varying the thickness of the connecting arrows. The third shows the most important interactions, in terms of their impact on community structure (as measured by their effects on the abundance of other species).

Many food webs have been generated, and several general patterns have emerged, with plenty of theories put forward to explain them. Many ecologists now believe, however, that some of these patterns may merely be artefacts of the imperfect data on which they are based. For example, some patterns seem to change depending on how intensively the community is sampled, and on the extent to which different species are lumped together (for pragmatic reasons) into a single point in the food web.

see also...

Community; Community structure; Ecological energetics; Trophic cascade(s); Trophic level(s)

Functional groups

A major focus of ecological studies has been the species, an approach that often stresses the uniqueness of individual species. An alternative approach – that of functional groups – emphasises the similarities between unrelated species by lumping together species that share structural or ecological characteristics. For example, the abundance of various species of algae on coral reefs varies from year to year in a very unpredictable way, however, when the species are placed into functional groups then the changing patterns become much more predictable.

The functional group approach attempts to reduce the complexity of ecosystems into manageable and ecologically meaningful chunks. Such an approach might be especially useful when making comparisons between communities in different geographical areas, or separated by long time periods, as although the species fulfilling the roles are different, the functions remain the same.

There is no single basis on which functional groups are constructed,

they might be based on growth form, behaviour, life history traits, some functional role (e.g. decomposers), or on their response to environmental factors (e.g. drought); which is most useful will depend on what is being studied. Sometimes size or shape is a useful way of grouping species, for example, trees, shrubs, herbs and grasses, however, sometimes life history attributes (seasonality, seed size and number etc.) might be used.

Functional groups are similar to guilds in that they both group species together based on functional rather than taxonomic similarity. They differ from guilds, however, in that the latter are specifically based on similarities in resource use, although the two terms have sometimes been used synonymously.

see also...

Community structure; Ecological redundancy; Guilds

Gaia

Gaia, is a difficult concept to pin down. In its original formulation by James Lovelock over 30 years ago, was a quasi-mystical idea that the Earth is a superorganism that actively, almost purposely, regulates itself to keep environmental conditions optimal for life. An intermediate view is that living organisms have a strong regulating effect on the global environment, without the connotation that this regulation leads to optimal conditions for life. Finally, there's the view that life has important effects on physical and chemical processes without having a regulatory role.

Scientists' views range from completely ignoring the idea, to cautious acceptance of some of the ideas, to the view that sees Gaia as explicable in terms of more conventional ecology. Few scientists subscribe to the superorganism view in which the Earth actively keeps the environment optimal for life. From our human point of view the world does, seem optimal, it's not too hot or too cold, and there's the 'right' amount of oxygen in the atmosphere. Life, however, evolved in a largely oxygen-free atmosphere so for the descendants of these organisms the present-day atmosphere is far from optimal, it's downright poisonous.

A key stumbling block to broader acceptance of the intermediate view is that it's at odds with mainstream evolutionary thinking in which natural selection operates at the level of the gene. In an attempt to counter this 'Daisyworld' was created. In this simple model, global temperature can be regulated by the interactions and feedbacks between black, white and grey daisies (the white daisies reflect heat, black daisies absorb heat) without invoking any superorganismic tendencies.

Whether or not the Gaia hypothesis, in whatever version, is found to be true is irrelevant, the general idea has been immensely important because it has stimulated a great deal of debate and has lead to fresh avenues of research. It's made biologists, geologists, oceanographers and atmospheric scientists talk to each other, and to look at the familiar in a new light.

see also...

Balance of nature

Generality in ecology

Science is all about finding general patterns among the diversity of nature, and seeking explanations for these patterns. A continuing question in ecology is how general the patterns are, and how generally applicable the theories put forward to explain them are. It seems that there are no 'universal' ecological laws that apply over all spatial and temporal scales; all patterns, and the processes that underlie them, are contingent to varying degrees. Contingent means that laws and theories apply only under particular conditions.

There are certainly general, widespread ecological processes (competition, disturbance, predation, etc.) but their importance varies between systems in ways that aren't predictable. They are contingent on the organisms present, and on the environment in which the interactions are taking place. History and location are important generators of contingent patterns. For example, biomes are general patterns that are contingent on fire and soil conditions. These latter factors aren't merely details that partially obscure the general pattern, in some cases their effects override the influence of climate.

In the case of biomes the contingency is manageable. Given this combination of rainfall and temperature, under this fire regime on this particular type of soil ecologists can make a good guess at the dominant sort of vegetation they would find there. Contingency isn't so manageable in other areas of ecology. Part of the difficulty with ecology, particularly community ecology, is that it deals with 'middle-number systems'. It's quite easy to make predictions about the behaviour of a few objects, likewise it's quite easy to make predictions about the behaviour of a large number of entities by using averages. The problem comes when you're dealing with entities that interact (often indirectly), and are too numerous to analyse individually, but too few to allow statistical averaging.

see also...

Biomes; Macroecology

Global environmental change

The Earth is undergoing a number of human-induced environmental changes, in climate, in atmospheric composition, in land use, and declining biodiversity. Underlying these changes is human population, which since 1900 has increased *three-fold*. We now use 30 times more fossil fuel and have increased industrial output by 50 times over the last 100 years.

Looking at just one facet of global change in isolation shows how dramatic the consequences are likely to be. Largely as a result of increased emissions of CO_2 and other 'greenhouse' gases, climate is expected to change, particularly at high latitudes. In addition, deserts are likely to expand and at the same time global rainfall is likely to increase; mass migrations of refugees from droughts or floods can be expected; there will be more extreme weather patterns; and sea level is expected to rise by 50 cm over the next 100 years.

Responses to the increase in temperature have already been noted in many species. The ranges of 22 species of non-migratory European butterflies have shifted northwards by up to 240 km over the last century, and the ranges of 59 species of bird have shifted by an average of over 18 km over the last decade; several birds and amphibians are also breeding earlier. What's more, these changes have occurred at warming levels of less than a fifth of those expected over the next 100 years.

Over the next century land-use changes will have the biggest impact on terrestrial biodiversity, followed by climate change and nitrogen deposition. Impacts are likely to vary between biomes, with Mediterranean and grassland biomes perhaps suffering the largest proportional loss in biodiversity. As with all future global-change scenarios, however, there remains a huge amount of uncertainty, particularly at regional level, arising largely from an unknown degree of interaction between the various factors. Although the future is uncertain, it is clear that global change is occurring now, and will continue for the foreseeable future, probably at an accelerating rate.

see also...

Biodiversity; Biomes

Grasslands

Most natural temperate grasslands (prairies, steppes, pampas) are found in the interior of continents, where it's too dry for forests and too wet for deserts. In areas that aren't too dry for forests, grazing and fires can maintain areas of grassland. Virtually all the natural temperate grasslands were, until recently, heavily grazed by large mammals (up to 60 million bison once roamed the plains of North America).

Winters are cold or mild and the summers hot, making grasslands prone to fires. Temperate grasslands have some of the world's most fertile soils, and huge areas have been converted to agriculture.

To understand the ecology of grasslands we distinguish between natural, semi-natural and man-made grasslands. Natural grasslands were formed, and are maintained, by the interplay of climate, soil factors, grazing and fire. Semi-natural grasslands were formed and modified by humans, but the plants haven't been sown. The grasslands resulting from the clearance of woodlands in Western Europe are a good example; if these areas were left unmanaged they would revert to woodland.

So where did the plants now associated with semi-natural grassland live before humans created their habitats? There were a few pockets of natural grassland at high altitudes, or on very thin or infertile soil, while other species probably grew in gaps within the woodland. Some semi-natural grasslands have a very diverse flora and are therefore considered to be of high conservation value and are actively managed to prevent succession to woodland.

The vast majority of the plant, fungus and invertebrate biomass in temperate grassland is underground. Here, the dense mass of roots is suffused with symbiotic fungi, and together they make up the mycorrhizal network. This provides a rich food source for immense numbers of invertebrates.

see also...

Biomes; Savannah; Symbiosis

Guilds

It's rarely possible to study entire communities; there are simply too many species. Ecologists therefore often divide communities into more conspicuous, more manageable, but hopefully still ecologically relevant, bits. Examples include trophic levels, functional groups, and guilds.

In ecology a guild is 'a group of species that exploit the same class of environmental resource in a similar way ... without regard to taxonomic position, that overlap significantly in their niche requirements' (Root, 1967). Note that there are no taxonomic restrictions on guild membership, the focus is on how species use a resource, not on their taxonomy. For example, the various species of seed-eating ants and rodents in the Arizona desert constitute a guild. Having said that, guilds often do seem to consist of taxonomically related species, though the extent to which this is a reflection of nature, or merely the taxonomic bias of ecologists, is unclear.

As with most ecological definitions, there is a lot of confusion and multiple interpretations of the concept. For example, phrases like 'same class of' and 'in a similar way' are open to interpretation. Its relationship to the similar/parallel/synonymous idea of functional groups is also unclear.

When studying competition between species it makes intuitive sense to focus on guilds; because guild members use the same resource in a similar way they are likely to show the most intense competition. This doesn't mean, however that competition is inevitable between species sharing a resource, for this to happen the resource must be a 'limiting factor'. For example, the population sizes of many herbivorous insects are kept below the level at which they would compete. Neither does it mean that competition occurs only between guild members; there are other types of competition not involving competition for resources.

see also...

Community; Functional groups; Interspecific competition; Limiting factors; Niche

Habitat

A habitat is the place where an organism lives, for example, a stream, a rocky shore, or a woodland. While this might be specific enough for larger animals and plants, for smaller organisms more detail would be useful. In a woodland for example, invertebrates occupy habitats, in the soil, in deadwood, in dung, in the tree canopy etc. Habitats are therefore best seen from the organism's point of view. At some arbitrary point along the continuum habitats become microhabitats, for example the larva of the holly leaf miner (a fly) lives inside a single holly leaf.

The diversity of habitats in a geographical region is often correlated with species diversity, in that areas with greater habitat diversity tend to have more species. The architectural complexity of a habitat is also important. For example, a closely cropped grassland would be expected to support fewer species of aboveground insects than grassland containing a range of plant heights and structures.

When considering the geographical range of a particular species, the habitats near the edge of the range tend to be less suitable than are those at the centre. Even within a relatively small geographical area habitats often vary in quality, with populations in some poorer habitats possibly being maintained by immigration of 'surplus' individuals from high-quality habitats, here habitat quality is measured in terms of the population's growth rate. This so-called source-sink dynamics has important implications for managing habitats for conservation.

One current concern is the widespread destruction and degradation of habitats, indeed habitat destruction is thought to be one of the most important threats to global biodiversity. One of the effects of habitat destruction is that the remaining habitats become very fragmented, that is they become smaller and more isolated. The worry is that populations occupying recently fragmented habitats face increased risk of extinction.

see also...

Habitat fragmentation; Landscape ecology

Habitat fragmentation

In the 1500s over 80% of the state of São Paulo, Brazil, was covered with rainforest. Five hundred years on, there's little left. Where other habitats existed within a forested matrix, there are now a few small 'island' forests in a sea of development. This story is repeated all over the world; indeed, the fragmentation of natural habitats is one of the main threats to biodiversity.

Fragmentation leads to a reduction in the *amount* of habitat (hence to a reduction in biodiversity), and also increases *isolation* of remaining fragments, further increasing risk of extinction.

Fragmentation leads to physical and biological 'edge effects', resulting from the creation of an edge at the boundary of two habitats. For example, the edge of a forest is significantly warmer, lighter and less humid than the 'core', and there may be an influx of commoner organisms from the surrounding matrix.

An important implication of this is that habitat fragments (or nature reserves) below a certain size will essentially be all edge and no core, and may no longer be able to sustain populations of core habitat specialists.

One possible solution to offset the isolation aspect of fragmentation is the maintenance (or construction) of 'wildlife corridors' to aid movement between isolated habitats. But: the corridor needs to be wide enough to contain some core habitat, otherwise core species won't use it. The species needs to be sufficiently mobile to use the corridor (and many rare species aren't good at dispersing). Corridors also carry risks; they might allow 'undesirable' elements such as predators or diseases to move between fragments.

So, do we focus our limited resources on maintaining corridors, even though the evidence for their use as conduits is limited? Or do we focus on mitigating (often untested) edge effects? As with most conservation issues, there is no simple answer.

see also...

Biodiversity; Island biogeography theory; Landscape ecology; Metapopulation; Species/area relationship

Habitat management

'**O**ur perceived need to manage habitats or species is an admission of failure' (Sutherland, 1998).

Many habitats and species of conservation importance, at least in Europe, are characteristic of the early stages of succession. In prehistoric times such habitats would be maintained over the landscape level by natural disturbances creating suitable habitats, ranging from a tree falling over, to floods, to 'stand-replacing' fires. This isn't an option in today's highly fragmented landscape, hence the need to manage existing small patches of isolated habitat.

The key concept underlying most habitat management is succession and trying to slow it down (whereas most habitat creation involves kick-starting succession). This usually involves creating disturbances of some kind, often by grazing, cutting or burning.

Many myths, surround habitat management which, when subjected to testing, are often found to be wanting. For example, the management of ponds: big or deeper ponds are better, the temporary drying out of a pond is disastrous and so on.

With habitat management whatever you do will benefit some species but be damaging to others. The starting point of any management is therefore having clear goals in mind: what is the habitat to be managed for, and why? These are essentially value judgements. It's also crucial to know what species are present in a habitat *before* it is managed. One way of managing larger sites is to do so on a rotational basis, thereby creating a mosaic of habitats.

Finally, habitat management needs to be seen in the context of the broader landscape, particularly when habitats are small and isolated. What goes on outside the habitat is often as important as what goes on inside.

see also...

Conservation; Habitat (re)creation; Landscape ecology; Restoration ecology; Succession

Habitat (re)creation

In industrialised countries most fuel and building materials come out of the ground, and in order to get at them habitats are destroyed. Once the extraction has ceased you're usually left with a hole in the ground and/or a pile of (often toxic) waste material. In addition, activities such as road building, and the planting of agricultural monocultures and biomonotonous forestry plantations contribute to a decline in the amount of natural and semi-natural habitats.

Starting with a man-made habitat, or even a piece of bare ground, it is now possible to create semi-natural habitats. The question is, should we? Some people argue that habitat creation gives industry a 'licence to kill'. In other words, it doesn't matter that a habitat is destroyed because we can simply create a new one on the site after the extraction has ceased, or in a different location in part-exchange for the destroyed area. Although there have been successes, many attempts at habitat creation ultimately fail, partly because newly created habitats need continued management and there are questions about the ability of developers to deliver this sort of long-term commitment.

Of course a newly created habitat can never be an exact replica of what was destroyed, as each habitat is the product of a unique history that is intimately linked to the landscape in which it developed. Given a choice, conservation is always preferable to habitat creation.

Having said all this, industrial development has already destroyed many habitats, leaving behind a legacy of post-industrial dereliction. Habitat creation is clearly beneficial in these circumstances, although this is not always the case: a few post-industrial sites in the UK are now nature reserves because of the rich and unusual plant communities they support. As these activities will continue for the foreseeable future, habitat creation will be an important tool with which ecologists can mitigate some, but never all, of the destruction.

see also...

Conservation; Grasslands; Habitat management; Restoration ecology; Succession

Herbivory

Herbivory means eating plants. This term hides a lot of variation in the nature of the interaction between herbivores and plants. Sometimes herbivory means predation, where the whole plant is killed and eaten. Aphids are parasites of plants. Sheep are grazers; they eat bits of several plants and don't usually kill them, at least not those that are adapted to being grazed.

The impact of herbivores on plants is very complex; it depends on which part of the life cycle is eaten, which bits are attacked, how much, when, and how often. Most studies on herbivory have focused on what goes on above ground, but the underground parts of plants are also eaten.

Plants have two basic responses to being eaten: they can try to avoid it, or they can tolerate it. Avoidance often comes in the form of chemical, physical (thorns etc.), or other defences (e.g. ants protecting acacia trees from herbivores). Tolerance is the ability of plants to regrow or reproduce after being eaten. For example, in times of plenty, perennial plants often store carbohydrates, which they may call upon to compensate for losses arising from herbivory. Tolerance might therefore be expected to be more common where nutrients are more plentiful to support regrowth (although sometimes the opposite is found to be the case). Likewise, it might be expected that slower growing plants would find it more difficult to replace lost tissue, making tolerance less likely; such plants might be expected to invest more in chemical and physical defences.

Some plants can overcompensate for the effects of moderate levels of herbivory, in that plants subjected to herbivory grow at a faster rate than those that aren't grazed. Whether some plants actually *benefit* (in terms of their fitness – see Behavioural ecology, p. 3) from being eaten is an intriguing, complex and controversial idea.

see also...

Chemical ecology; Predation; Top-down/bottom-up

Historical ecology

You can't fully understand the ecology of the present without knowledge of the past. Historical ecology is the 'history of vegetation and landscape' (Rackham, 1998). Most habitats have been affected by humans over thousands of years; most landscapes are to varying degrees cultural or semi-natural. The distinction between natural and semi-natural landscapes is becoming increasingly fuzzy.

For instance, the use of fire by native peoples has had major impacts on landscapes, from the Australian bush to the Scottish moors. The arrival of humans in the northern forests of America coincided with the extinction of several large mammals, such as sabre-toothed tigers, giant beavers and mammoths, although the extent to which the extinctions were caused by hunting, climate change or some combination of these will never be known for sure. American Indians also burnt and cleared patches of the forest. When Europeans arrived, they thought they were entering a primeval wilderness, but they were mistaken. The time scale of historical ecology ranges from decades to millennia. In the Northern hemisphere, the time since the last glaciation is often considered to be a useful starting point.

Historical ecologists use a variety of techniques to learn more about past communities. For instance pollen buried in layers at the bottom of lakes or in peat bogs can tell us how vegetation has changed over time. Ancient tree rings can tell us which years were good for growth and when fires occurred. Then there are maps, documents, and aerial photographs.

An historical perspective, although often neglected, is important when trying to restore or manage habitats. What this perspective often indicates is that there is no single natural state and no single reference point in time on which to base our management or restoration attempts.

see also...

Habitat management; Habitat (re)creation; Restoration ecology; Scale in ecology

48

Indirect effects

In the Arizona desert rodents and ants compete for a limited supply of seeds. So when rodents were excluded from certain areas using rodent-proof fences, it came as no surprise that the number of ant colonies soon increased. After a while, however, ant populations began to decline in the rodent-free zones. Why?

Rodents eat a range of seed sizes, but prefer larger seeds. So, in their absence, large-seeded plants began to increase in abundance, which then outcompeted small-seeded plants, reducing the preferred food supply of the ants.

This is an indirect effect, which occurs when one species affects the population size of another via the effects on a third, intermediary, species. In contrast, a direct effect involves a physical or chemical interaction between two species. Indirect effects can make the interpretation of field experiments difficult. In Arizona the effects were only noticed after a few years; few field experiments are conducted over this length of time.

Indirect effects can result from a chain of direct effects (e.g. trophic cascades), or they can involve one species affecting the interaction between two other species (e.g. predator-mediated coexistence). Indirect effects can be negative, as in some forms of interspecific competition, or positive, as is the case with indirect mutualism ('my enemy's enemy is my friend').

Indirect effects appear to be widespread, but as few comparative studies have been done it is difficult to make generalisations about their importance compared with direct effects, let alone how frequent and under what conditions the different types of indirect effect occur. It seems likely that strong direct effects are needed to produce noticeable indirect effects, but a lot more experiments are needed to test ideas like this.

see also...

Interspecific competition; Predator-mediated coexistence; Species introductions; Trophic cascade(s)

Interspecific competition

The prevalence and importance of interspecific competition has been one of the most hotly contested issues of ecology.

Interspecific competition simply defined is a detrimental interaction between two or more species (see Species interactions, p. 90). Often the interaction is highly asymmetrical, with one species suffering much more than the other. There are several ways in which these reciprocal negative effects can arise, ranging from indirect interactions such as competition for a resource in limited supply (exploitation competition) or sharing a predator (apparent competition), to direct interactions such as direct physical or chemical interference between organisms to prevent access to a resource (interference competition). An example of interference competition is provided by barnacles. Space is at a premium on rocky shores, and barnacles aren't averse to bulldozing their neighbours off the rock.

Darwin argued that interspecific competition would be strongest among closely related species, as these would tend to have similar resource requirements. Although competition has since been found between distantly related species, Darwin's idea is still sound.

Ideas about the importance of competition have been refined over the years. First it was thought to be very common and important, then several ecologists championed the importance of predation or disturbance in structuring communities. Ecologists then recognised that competition was important in some groups of organisms (e.g. plants) but not others (e.g. herbivorous insects). More recently, it's been found that competition is actually quite widespread among herbivorous insects, apart from free-living insects that chew leaves.

There are basically two outcomes of competition: either one species eliminates the other (competitive exclusion), or they coexist.

see also...

Character displacement; Guilds; Niche; Resource partitioning; Species coexistence; Species interactions

Intraspecific competition

Individuals belonging to the same species have similar requirements. If there is not enough of a particular resource to satisfy the needs of the whole population then there will be competition between individuals for access to that resource. The resource could be food, space, light; anything that is 'consumed' in some way.

The effects of competition between individuals of the same species (intraspecific competition) are density dependent, i.e. the higher the density the greater the impact of competition on each individual. Indeed intraspecific competition is considered to be one of the main processes that act as a brake on population growth. Some populations may never reach a high enough level, or may not deplete their resources sufficiently, for competition to be important.

Competition can involve direct aggression (interference competition), which may be physical, psychological or chemical. For example, males competing for females may fight, engage in displays to ward off other males, or use scent to keep other males away. Competition for mates, space and light often involves interference competition.

Competition isn't always so direct. The fact that part of a resource is eaten by one individual means that it is no longer available to others. In this case competition is indirect, the effects being experienced via the depletion of the resource. This is called exploitation competition. There are probably elements of both exploitation and interference in most competitive interactions.

Often competition is asymmetrical, with some individuals suffering much more than others. The ultimate effect of competition is that individuals have a lowered fitness, i.e. their relative genetic contribution to the next generation is reduced. This could be because they die, or fail to breed, or grow up to be smaller than they otherwise would have been.

see also...

Density dependence; Limiting factors; Population regulation

Island biogeography theory

Want to find out how the size of an island affects the number of species living on it? Easy: chop some in half to see what difference it makes. Want to know how immigration rate is related to distance from the mainland? Just cover the islands with huge tents, kill all the insects inside, and then wait for new immigrants to arrive.

The islands were mangrove islands and the experiments were designed to test the 'equilibrium theory of island biogeography'. The central tenet of the theory is that the equilibrium number of species on an island (see Equilibrium p. 33) is determined by the balance between extinction and immigration rates. This equilibrium is dynamic: individual species come and go, but the overall number stays roughly constant.

The theory offered a biological explanation for one of the most widespread patterns in nature: the species/areas relationship (the bigger the area, the more species it contains).

It was also used to provide some 'rules' for the design of nature reserves. We've since learnt that you can't apply simple, general theories to such complex issues.

The theory of island biogeography and the evidence used to support it have been criticised on several grounds, and its influence in conservation biology has since waned, to be replaced by the rising star of metapopulation theory. The theory has, however, played an important role in the development of ecology. It helped ecologists to recognise the importance of landscape level processes and, more importantly, it stimulated a huge amount of research that has increased our understanding of the natural world.

see also...

Habitat fragmentation; Landscape ecology; Metapopulation; Models in ecology; Species/area relationship

Keystone species

In one of the most famous experiments in ecology, Robert Paine removed all specimens of starfish *Pisaster* from an area of rocky shore, and monitored the effects on the community. The result was clear: in the absence of *Pisaster*, the number of species nearly halved. This was because mussels competed strongly for available space and exluded other species.

In contrast, when *Pisaster* was present, it predated heavily on the mussels and held numbers in check, preventing them from monopolising the space. Clearly *Pisaster* played a vital role in maintaining the diversity of this rocky shore community, analogous to the keystone that stops bridges collapsing. Hence the term keystone species.

Since then the concept has been both broadened and refined. It now refers to a species 'whose impacts on its community or ecosystem are large, and much larger than would be expected from its abundance' (Power and Mills, 1999).

Some conservationists have argued that the keystone concept could be very important in prioritising conservation effort. The idea is that, with insufficient time and money available to conserve all species, you should concentrate your efforts on keystone species, as this should help maintain species diversity. Conversely, if the keystone species goes extinct then a lot of other species will follow.

Although this is an appealing idea, there are dangers in applying the concept to determine conservation priorities. This stems from the difficulty identifying the keystone species (if any) in particular communities. Keystone species are context dependent (even *Pisaster* isn't a keystone species on other rocky shores). So, although an important ecological concept, its application to conservation is, for the moment, limited.

see also...

Ecological redundancy; Ecosystem engineers; Trophic cascade(s)

Lakes

Most lakes are found in the Northern hemisphere, in areas subjected to glacial activity. Such lakes are therefore less than 12,000 years old. Exceptions include the huge, ancient lakes formed by tectonic activity, such as Lake Baikal and Lake Tanganyika.

The eventual fate of lakes is to be filled in by sediment eroded from upstream or from the surrounding land; they are therefore more temporary features than rivers. Shallow ponds fill up the quickest, partly because there's less of a hole to be filled, but also because vegetation growing around the pond's edge speeds up the process by adding plant detritus.

Deeper lakes in temperate regions undergo 'thermal stratification' each spring. The result of this process is a warm layer of water overlying a much cooler layer; the transition zone between the two layers is characterised by a marked decrease in temperature and oxygen concentration.

Primary production in lakes is dominated by photosynthesising plankton (lakes are generally too deep to support rooted plants, except around the edges). Due to the limited penetration of sunlight, these plankton are confined to the upper layers. As their populations increase over the summer they may use up all the available nutrients, and when they eventually sink to the bottom the nutrients don't return to the upper layer because the upper and lower layers of the lake don't mix.

In addition, unlike the sea floor, the bottom of most lakes has few species, largely because the water becomes oxygen depleted in the summer as bacteria use up oxygen when decomposing plankton, and oxygen isn't replenished from the upper layer.

In winter, stratification breaks down and the whole lake mixes. Whether or not stratification occurs, and how frequently, depends on latitude. The process also occurs in the oceans; indeed oceans and lakes have many similarities.

see also...

Oceans

Landscape ecology

Landscape ecology is a new discipline that looks at the bigger picture. It deals with the spatial patterns, processes and changes occurring at the scale of hectares to square kilometres.

Areas of this sort of size comprise a number of patches, which include distinct ecosystems (e.g. streams, woodlands), successional stages and different land uses (e.g. agricultural land, urban areas). The interaction aspect is important; what happens to one patch is likely to affect adjacent patches. For example, the quantity and quality of water entering rivers depends on land use within the catchment area, and what goes on in habitats adjacent to a nature reserve may be more important than what goes on *within* the reserve.

The patterns that are measured at the landscape level include size, shape, 'connectivity', and position of patches in relation to each other. The processes studied range from nutrient movement to animal dispersal. The other major aspect involves trying to understand the causes and consequences of changes to the landscape over time, including such things as habitat fragmentation, disturbance (e.g. fire), successional and climatic changes.

The scale looked at should depend on the questions being asked and the organisms being studied. Having said that, most landscape ecology involves the study of large scales, requiring the use of new techniques such as computer-based geographical information systems (GIS) which enable the spatial pattern of landscapes to be measured accurately. The information can then be manipulated and used in conjunction with mathematical models to predict changes in landscape patterns and processes arising from various human impacts.

A host of important issues require a landscape-level perspective to understand them fully. The discipline of landscape ecology, though it still lacks a theoretical framework, will therefore come to play an increasingly important role.

see also...

Dispersal; Habitat fragmentation; Metapopulation; Scale in ecology

Latitudinal diversity gradient

One of the best known patterns in ecology is the general increase in the number of species found as you get nearer the Equator. This gradient in diversity is found in both terrestrial and aquatic systems, in vertebrate, invertebrate and plant groups. There are some exceptions – including parasitoid wasps, aphids, some marine groups, and some parasites of vertebrates – but the majority of taxonomic groups follow the pattern.

Now for the tricky bit: what's the underlying cause of this pattern? Ecologists are never short of ideas: at least 28 hypotheses have been put forward. These range from the increased energy availability in the tropics, to their large area, their greater stability, and their great age. Most of the explanations are consistent with some bits of evidence but not others, work for some groups, but not others.

A key problem is that the pattern is on such a large scale that we can't use experiments to test the hypotheses. All we have is sometimes messy observational evidence, which isn't very useful when trying to distinguish between hypotheses. And finding evidence in support of one theory doesn't rule out the operation of others. Also, it's rather optimistic to expect a single cause to be able to explain the gradient in taxonomic groups as different as trees and marine molluscs.

For some groups, the greater diversity in the tropics is apparently quite easy to explain. For example, the great diversity of tree species in tropical forests means that there are always some trees in fruit; it isn't therefore surprising to find a high diversity of fruit-eating birds such as parrots in the tropics. Outside the tropics they'd either starve or have to migrate in winter.

Despite knowing of its existence for decades, there is still little consensus about the processes generating one of the most prevalent patterns of life on earth.

see also...

Biodiversity; Macroecology; Species/area relationship

Life forms

Unencumbered by any botanical knowledge, anyone could probably describe the general appearance of most of the world's biomes: tropical rainforests are full of trees; grasslands have few or no trees but lots of grass; and deserts have lots of space. There is also a remarkable similarity in the general appearance of the same biome in different continents.

The reason for these similarities in 'life form' of the plants is the similarity of climate within particular biomes. In their response to the prevailing climatic factors, different plants have evolved similar strategies.

It was the Danish botanist, Christen Raunkiaer, who devised an elegantly simple classification of plant life forms, based on the ways in which plants protect their buds (the parts that produce new shoots) during the most unfavourable times of the year. Raunkiaer recognised five basic types of plant life forms which, in various combinations, go to make up each biome's flora.

In the warm, wet tropics, trees are by far the commonest life form. Under such ideal growing conditions, light becomes an important factor, and so it pays to be tall. The buds don't need protection from the elements and so are freely exposed. In forests with cold or dry seasons scales protect the buds.

In higher latitudes subjected to periods of extreme cold (e.g. tundra) a lot of plants hold their buds close to – but not under – the ground where it's warmer (below ground the soil is frozen in winter). Being covered by relatively warm snow gives additional protection from the intense cold.

In temperate regions many plants die back to ground level, where the bud is protected during the winter by a covering of dead leaves and soil.

In deserts and other arid regions there is an abundance of annual plants, which avoid drought by lying dormant as seeds until the rain comes. The other main life form in these biomes is plants that survive (avoid) the drought as underground tubers.

see also...

Biomes

Life history strategies

Each organism can take in a limited amount of energy and nutrients in its lifetime, so the way in which it allocates these resources becomes crucial. For example, there is a cost to reproduction. Energy invested in reproducing can't be used for other things; the cost is reduced growth or survival. Sometimes individuals delay reproducing and put their resources into growth. In forests, for instance, it's important for trees to reach the canopy, and sunlight, to ensure their survival; so they put a lot of resources into growth, only reproducing once they've reached the canopy.

An organism's life history is its pattern of growth and reproduction throughout its lifetime. In most habitats there isn't an optimal life history strategy, the best strategy is one that maximises an individual's fitness (see Behavioural ecology, p. 3), within the constraints set by its evolutionary history, and on what individuals of the same and other species are doing.

Some species complete their life cycle within a year, others live and reproduce repeatedly over a number of years, and still others live for a number of years, breed once and then die. Some species produce many offspring, others invest in a few young, which they usually care for. Although very different, all these strategies work, at least under certain conditions. Finding out under what conditions, and the trade-offs involved is an important area of study in ecology.

In unpredictable or short-lived habitats small 'weedy' species with high reproductive rates and good dispersal abilities tend to predominate. These species are able to find the habitat quickly, breed, and leave before it changes or disappears. However, these species tend to be poor competitors and usually lose out to slower growing, bigger, better competitors that eventually arrive. As long as new habitats appear frequently enough then both extremes of the strategic continuum work.

see also...

Life forms; Species coexistence

Limiting factors

The idea of limiting factors has been used in agriculture for some time. A shortage of nutrients like nitrates and phosphates may limit the productivity of crops, so adding these nutrients will increase the crop yield. In arid regions it might be water supply that limits crop yield, so adding water will likewise increase the crop yield. Here the limiting factor is the resource that is in the shortest supply relative to the needs of the plant for growth.

Where populations are concerned, a factor is described as limiting if a change in it produces a change in the average density of a population. For example, nesting site availability can be shown to be a limiting factor for a bird population if adding nesting boxes increases its size. In one experiment it was found that shooting woodpigeons had no impact on population size. This was because food supply was the factor limiting the size of the population, shooting the birds meant there was more food for survivors, and immigrating pigeons also supplemented the population. This is the principle behind managing game birds such as red grouse.

More than one limiting factor probably operates at any given time (or sequentially throughout the year), and they are likely to interact to determine the size of a population.

It is important to distinguish between the factors that regulate populations and those that determine their average abundance. Only density-dependent factors can regulate population size (i.e. keep it within bounds), whereas both density-dependent and density-independent factors determine the average population size.

An understanding of limiting factors is central to many areas of ecology, ranging from interspecific competition, to the control of pests, to predicting the impact of rising levels of CO_2 on plant productivity.

see also...

Density dependence; Population regulation; Top-down/bottom-up

Macroecology

Over the last decade macroecology has become an increasingly popular approach to studying ecology. Whereas the majority of ecologists look in detail at the subtleties and complexities of interactions between species over small spatial and temporal scales, macroecologists take a much broader view.

The effects of some ecological processes are only seen over large scales, so they aren't always amenable to experiments. Another approach is therefore needed. One approach is to look for large-scale patterns in nature, and then to seek explanations for them – this is the essence of macroecology.

Showing that the patterns are real isn't simple. For patterns to emerge from the biases and vagaries of the data large samples are needed, which means that the better known groups of species form the bulk of the studies. However, if general widespread patterns are discernible, then it suggests that general widespread ecological processes might underlie them. General patterns that have been found include the latitudinal diversity gradient, the species/area relationship, and the links between body size, abundance and geographical range.

The main problem is uncovering the causes behind the patterns. Without an experimental approach it's not easy to distinguish between them. What's more, many of the patterns are probably caused by more than one mechanism, so determining their relative importance will be difficult.

The lack of experimental rigour has been a major criticism of the macroecological approach. However, a large-scale approach to ecology is still necessary. Many of the criticisms levelled at macroecology also apply to the fossil record, yet where would our understanding of evolution be without that?

see also...

Experimental ecology; Generality in ecology; Latitudinal diversity gradient; Scale in ecology; Species/area relationship

Mediterranean shrubland

Mediterranean shrubland is found on the west coasts of continents at around 30–40° latitude. It includes the area around the Mediterranean, the Californian chaparral and the South African fynbos.

The climate of this biome is characterised by lots of rain in the mild winters and drought in the hot summers. Summer fires play a pivotal role in the ecology of Mediterranean shrublands.

To cope with the summer drought most plants have an extensive, shallow network of roots plus a deep root that can reach tens of metres deep in some species, even extending into the bedrock. The vegetation is dominated by evergreen shrubs with thick waxy cuticles to reduce water loss; their intertwining branches forming impenetrable thickets. Seedling establishment or resprouting from the underground rootstock is difficult under these conditions, unless gaps open up in the canopy. This is where fire comes in: not only does it release nutrients into the soil, it also creates gaps in the vegetation.

Many plant species either have a seed bank that lies dormant until stimulated by smoke or heat, or fire-resistant rootstock that can quickly resprout after fire. Indeed such is the dependence on fire that it has been suggested that some species of plant have actually evolved increased flammability, thereby increasing the likelihood of fires. Fire-enhancing traits include dense canopies of fine branches containing a lot of dead wood, loose flaky bark and high volatile oil content in the leaves. All these traits have other equally plausible explanations, which is one reason why the hypothesis is still debated.

Why would a plant evolve to burn anyway, particularly if it kills itself in the process? One answer is that under certain conditions 'torch' plants could benefit if the fire kills the neighbouring plants ('damps'), leaving gaps and nutrients for its own offspring lying dormant in the seed bank.

see also...
Fire

Metapopulation

The natural world is patchy. And very often, patches of habitat suitable for a species to live in are separated by unsuitable habitats. Thus, at a regional level, a population often consists of a patchwork of discrete local populations: a 'population of populations'.

If there is migration of individuals between these discrete local populations, we have a metapopulation. The focus on migration is the key difference between the metapopulation approach and the traditional focus on births and deaths within a single breeding population. Within the regional metapopulation it is possible for individual local populations to go extinct, only to flicker back to life as they are recolonised by individuals from populations nearby.

Metapopulation theory is emerging as the main theoretical paradigm for conservation biology. Increasingly conservation strategies for single species are based on mathematical models derived from metapopulation theory.

Metapopulation ecology has a number of potentially important messages for conservation. To take just one example, the focus on managing isolated nature reserves is unlikely to be sufficient for those species that exist as metapopulations. These species need to be managed in the context of the surrounding landscape. This might involve, for example, actively maintaining habitat patches (currently) unoccupied by them.

There are also potential dangers if the metapopulation approach is applied blindly. For example, increasing habitat fragmentation has led to a lot of previously continuous populations becoming progressively more isolated, eventually resulting in an assemblage of *totally isolated* small populations, all declining towards extinction.

see also...

Dispersal; Habitat fragmentation; Island biogeography theory; Landscape ecology; Models in ecology; Patchiness; Population

Microbial ecology

Humans are impressed by size. Perhaps that's why, when we think of the Jurassic, we tend to think of the huge dinosaurs that 'ruled' the Earth. However, if anything can lay claim to ruling the Earth, it's the microscopic prokaryotes (mainly single-celled organisms that lack a nucleus, such as bacteria). They were here first, will probably outlast everything else, play an important role in just about every ecological process (from nutrient cycling to cloud formation), and are as near to being ubiquitous as makes no difference (they even live in rocks a few kilometres under the Earth's surface). And if it's size that impresses, then, in a genetic sense, prokaryotes are the largest organisms. A new strain of a disease-causing bacterium, arising from a single cell (hence it could be thought of as a single genetic individual) can engulf entire continents.

Microbial ecology is the study of the interactions between micro-organisms (prokaryotes, protozoa, some fungi) and their environment. As a subject it's been relatively isolated from 'mainstream' ecology and has made little contribution to ecological theory, although things are changing.

Studying the ecology of micro-organisms is far from easy, largely because of limitations in technology. A gramme of soil may hold several billion bacteria of an unknown number of species, of which we can culture only 1% in the lab; so you can begin to see some of the difficulties. (Even the term species doesn't sit well with bacteria as genetic information can pass between quite distantly related groups). However, with recent advances in molecular biology we are beginning to get a handle on the functioning of these anonymous hordes, particularly in terms of nutrient cycling. If anything, micro-organisms are likely to be even more important than we currently think.

see also...

Biogeochemical cycling; Gaia; Molecular ecology; Microbial loop; Mutualism; Symbiosis

Microbial loop

Sometimes a method or a technique can profoundly alter the way we view things. For example, the traditional view of aquatic food chains is that photosynthesising plankton are at the base, these are then fed on by zooplankton, which in turn are eaten by fish. This view changed substantially when it was realised that microscopic plankton were not caught by traditional sampling techniques. These nano- and pico-plankton are between 0.2 and 20 micrometres in size (1 micrometre is a thousandth of a millimetre) and are now known to be the main primary producers in many of the oceans. In fact micro-organisms make up the bulk of the biomass in the oceans. However, they are much too small to be fed on directly by zooplankton, and are instead fed on by single-celled protozoans, which in turn are food for zooplankton.

It was also realised that a substantial proportion of the primary producers died without ever being grazed, and that a large fraction of the open-ocean community is dependent on the production of dissolved organic matter. In aquatic systems organic matter is divided on the basis of size, with anything under 45 micrometres in size being called dissolved organic matter (DOM). There are three main ways in which DOM is released from organisms. First, photosynthesising plankton exude or leak DOM, second, zooplankton are messy feeders and spill a lot of DOM when eating, and finally, it has been found that viruses are important as they cause infected bacterial cells to burst, releasing the cell's contents.

The DOM would be lost to the open-water community were it not shunted into the 'microbial loop' which runs alongside the traditional food chain. Up to half of the net primary production, in the form of DOM, is processed by the microbial loop. DOM is first taken up by bacteria, which are fed on by protozoans, which in turn are eaten by zooplankton, at which point the microbial loop rejoins the classic food chain.

see also...

Biogeochemical cycling; Decomposition; Ecological energetics; Food web; Microbial ecology; Trophic level(s)

Minimum viable population

Small populations face several dangers. These dangers interact, making things worse and, eventually, below a threshold known as a 'minimum viable population' size, the population is sucked into an 'extinction vortex'.

The dangers faced by small populations include genetic problems, and the effects of random variation in population processes (e.g. birth rates, unbalanced sex ratios). It's now thought that genetic problems aren't of immediate concern to small populations; they have more pressing concerns, the main one being 'environmental noise': unpredictable variations in environmental conditions. These range from frequent small events (a bit of bad weather), to infrequent 'catastrophes' (e.g. hurricanes).

A problem for ecologists is that rare species are difficult to study, and time is short. Therefore ecologists use mathematical models to estimate the chances of a population becoming extinct over a defined time period. This is known as population viability analysis (PVA). Using these models, ecologists can get some idea of the likely impact of different management options on the survival of the population and plan their management accordingly. The models perform quite well over the short term (a few years) but, as time goes on, and environmental variation increases, the models are likely to seriously underestimate extinction probabilities.

Extinction vortices are a population's 'death rattle'. What's more important is to find out the reasons *why* the population got to be rare in the first place; we need to treat the cause of the rarity rather than the consequences. It's the same for PVA: the starting point is a small population, it doesn't address how the population got to be in need of a PVA in the first place. We have no theory to guide us, the reasons why species become rare are complex and idiosyncratic, and even when we know the reasons there is often little we can do to eliminate them.

see also...

Conservation; Models in ecology; Rarity

Models in ecology

The role of models in ecology has always been contentious. As a crude caricature of the debate, modellers argue that without a grounding in theory ecologists would be running around collecting lots of information without anything to link it all together or to explain what's going on. Meanwhile some practitioners have little time for what they see as models that are either too simplistic to be of any practical use, or are too esoteric and with little grounding in ecological reality. There's some truth in both viewpoints.

It's not the closeness of the model's predictions to the real world that is most useful: they can appear to be right, but for the wrong reasons, there's often little way of knowing. What's much more useful is when you *don't* get a good match between theory and reality. It might be that one or more of the assumptions are invalid: Which ones? Why?

Like experiments, models can't be simultaneously realistic, precise and general; the best they can hope for is two out of three. There are two basic types of mathematical model: simple, general models and detailed simulation models. Each has a clearly defined role; the trouble comes when they are used inappropriately.

Simple general models help ecologists think in a rigorous, quantitative way, based on clearly articulated assumptions. They are caricatures of nature that hopefully capture the essence of what's going on. They are not meant to provide detailed predictions or to be applied to a particular species in a particular habitat.

Simulation models are detailed species-specific models. They require a lot of information about the biology of the species, enabling reasonably detailed predictions to be made. They are useful in areas such as pest control and species conservation.

see also...

Experimental ecology; Generality in ecology; Metapopulation; Minimum viable population

Molecular ecology

here are always cases in the press that the so-called 'look-alike' problem in wildlife trade: the sale of legal wildlife products being used as a cover for the selling of illegal products.

One way round the problem is to use molecular genetic techniques, such as DNA fingerprinting, to analyse DNA from the samples and so identify the species. In the case of whales, some of the meat on sale in Japan has been identified as belonging to protected humpback and fin whales.

These examples highlight the use of recent advances in molecular biology in answering (and generating) important ecological and conservation-related issues. Molecular analysis can also be used to assess the potential of translocating individuals from one area to another, perhaps to augment a declining population. In this case molecular analysis can help decide which populations are genetically closest to the population to be augmented. This can be important as bringing in individuals that are genetically quite different can lead to genetic problems. Related to this is the vital role molecular ecology will play in tracking the release of genetically modified organisms into the environment.

Molecular techniques are also proving useful in other areas of ecology. For example, they have been used to determine paternity in several species of supposedly monogamous birds. They've shown that extra-pair matings can be a frequent occurrence, with a sizeable proportion of the offspring belonging to another male. This has led to a number of studies addressing the risks and benefits to a female of having brief encounters with non-pair males, the strategies used by males to prevent extra-pair matings (e.g. guarding the female, repeated matings), and how the effort the male puts into raising the brood might vary with likelihood of paternity.

Mutualism

From a flowering plant's perspective, bees are winged penises, transferring pollen from the male parts of one flower to the female parts of another. In return for this service plants provide a reward, in the form of nectar. This is a classic example of mutualism, an interaction in which both species benefit.

There are fascinating examples of mutualisms: ants that protect acacias from herbivores and receive nectar and proteins in return; ants that farm fungus in 'leaf-gardens' and even protect the fungus from garden pests by using antibiotic-producing bacteria; cellulose-digesting bacteria found in the guts of termites and herbivores. Mutualisms can be direct – often involving nutrient or energy transfer or the dispersal of pollen or seeds – or indirect, involving intermediary species.

It's not always easy to measure, or even define, what benefit each species gets out of the interaction, either in terms of individual fitness or population size. In the case of lichens, the cyanobacteria and most of the algae can lead separate lives, so it's not clear how they benefit from the association with their fungal partner. One suggestion is that sometimes the interaction isn't mutualistic at all, fungus is simply exploiting the alga.

Ecologists now see mutualism as a form of mutual exploitation in which the benefits (usually) outweigh the costs. Like other interspecific interactions, the benefits and costs vary according to circumstances, making mutualistic interactions much more dynamic than was once thought. For example, the close association between plant roots and mycorrhizal fungi is often said to be mutualistic: the fungus providing mineral nutrients for the plant that, in return, provides carbohydrates for the fungus. This may well be true in nutrient-poor soils, but when there are enough minerals for the plant not to need the fungus, it becomes a drain on the plant's resources: the interaction becomes parasitic.

see also...

Indirect effects; Species interactions; Symbiosis; Trophic cascade(s)

Niche

There is no such thing as *the* ecological niche, there are at least three current, and quite different, uses of this abstract and enigmatic concept.

One early less formalised, use of the term niche was as a description of the functional role of a species in the community, the Eltonian niche.

Later, the concept of the niche as a 'multi-dimensional hypervolume' was born. Here each axis on a graph represents the use of a resource (e.g. food type) or tolerance to an environmental condition relevant to the life of a species (e.g. temperature). The niche represents a species' use of resources and its range of tolerances to environmental conditions within which it can survive and reproduce. This is the Hutchinsonian niche and is the concept most widely accepted today. Hutchinson divided the niche into 'fundamental' and 'realised' niches. The realised niche is a subset of the fundamental niche, and takes into account the restricting effects of other interacting species.

Although in an abstract sense the Hutchinsonian niche is useful, it doesn't make for easy, practical ecology. Following on from the Hutchinsonian niche came the more utilitarian idea of the 'resource utilisation function' (RUF), which concentrates on a single resource dimension (although it can be made multi-dimensional). This idea of the niche is the most focused, but also most useful, when studying exploitation competition between species. Unlike the Hutchinsonian niche, the RUF focuses on what actually happens between the tolerance limits, more specifically, on how resources are partitioned between competing species.

Each type of niche has its problems but each, as an abstract concept, is useful in providing a theoretical framework within which to study the ecology of species. Which one is most useful will depend on the questions being asked.

see also...

Interspecific competition; Habitat; Resource partitioning; Species coexistence

Oceans

They're big. Oceans cover 71% of the Earth's surface, hold 96% of the available water, and some bits are over 10 kilometres deep. Altogether they provide 99% of the living space on the planet.

They're wet. Water is a remarkable substance, quite unlike any other compound of similar size. It exists in all three states on Earth (solid, liquid and gas); it has a huge heat capacity, so the massive oceans experience little in the way of temperature fluctuations compared with terrestrial systems; and, luckily for organisms that live in it, the solid phase is less dense than the liquid (so ice floats on water).

They're blue. At least in less productive waters; more productive (usually coastal) waters are green as some of the blue light that's normally reflected is absorbed by dissolved organic matter.

They move. Water movement in the oceans is very complex, but there are some broad patterns. Prevailing winds and the spin of the Earth drive surface currents around the ocean basins; these currents have huge impacts on both regional and global climates. There are also layers of deep-ocean currents, unconnected to the movement of the surface layer, which move huge masses of water around the oceans. Besides horizontal movement there are also regions of vertical movement (e.g. in Arctic waters cold, dense water sinks to the bottom). The most productive areas of the sea are found where nutrient-rich deeper water moves into the upper layers (upwelling) where they are used by photosynthesising plankton.

They're salty. Most chemicals present in the oceans come from the weathering of rocks on land, which are then dumped in the sea by rivers. Chloride originated as gaseous hydrochloric acid spewed from volcanoes early in Earth's history. Why isn't the sea getting saltier? Many chemicals end up in sediments (up to several kilometres thick) covering the ocean floor. These eventually turn into rock and, through tectonic activity, often end up back on land, and the cycle continues.

see also...

Abyssal zone; Lakes

Organisms

Ecology often involves the study of individual organisms, but what is an individual? It's easy to recognise individuals in animal species. This is because most animals are 'unitary'; each is recognisable as a separate entity and follows a predetermined pattern of development through different life stages (e.g. egg – larva – pupa – adult), each with a fixed number of body parts.

It's less easy with modular organisms, where development is indeterminate: the exact form is strongly influenced by environmental conditions (think of the shapes of trees buffeted by strong winds). The adult life stage is often sessile; examples include animals (corals and sponges) as well as plants. These are the dominant life forms in many habitats and are often important 'ecosystem engineers'.

The structure of modular organisms is based on repeated units (modules). Some species have a repeated branched structure (trees, corals), while others, like grasses, often break into physiologically separate but genetically identical parts (clones). Clonal growth of plants enables these non-mobile organisms to sample their environment over a wide area. For example a single clone of aspen consists of 47,000 trees, occupying 43 hectares.

Modular organisms also have an internal age structure, with some bits of the same organism being much older than other bits; indeed in the case of trees the bulk of the organism is dead. They don't stop growing and they don't age; they generally die because something kills them. Modular organisms therefore include the oldest living things on the planet.

The organism has sometimes been used as an analogy for the way ant colonies, ecosystems, and even planet Earth operate. Almost all biologists reject such analogies, as they are not consistent with current evolutionary thinking.

see also...

Ecosystem engineers; Gaia

Parasitism

More than half the species on the planet are parasites, yet it's only recently that ecologists have started to study them in detail, and then mostly from the perspective of the host – parasites are just another problem to overcome. But let's look at parasitism from the parasite's perspective.

What are parasites? They are organisms that obtain resources from their host, having a detrimental effect on them in the process (though they don't always kill the host). They have an intimate association with their host (living on or in its body), and usually attack only one individual during their lifetime.

From the parasite's perspective, hosts are species-rich, patchy habitats with a limited lifespan. As with all patchy habitats, dispersal is a key process. For parasites living on the host's surface dispersal is relatively easy although, on the down side, it isn't as easy to obtain resources as it is when living inside the host and (for parasites of birds and mammals) it's not as warm and cosy. For internal parasites, many of the symptoms of disease (sneezing, coughing, diarrhoea etc.) are dispersal mechanisms.

A good place to live inside animals is in the alimentary canal: plenty of food and convenient entry and exit points. Once outside the host's gut, however, it's not easy to get back in. The solution often involves using other species as intermediate hosts in what sometimes seems to be unnecessarily complicated life cycles. Some parasites even manipulate their host's behaviour for their own benefit.

The other main problem faced by parasites is that their habitats fight back; parasites elicit defensive responses from their host. It's often thought that parasites evolve to become less virulent so as not to destroy their habitat some parasites need to kill their host in order to be transmitted, others don't. The level of virulence attained will be that which maximises the parasite's fitness within the constraints set by its host.

> ### see also...
> *Coevolution; Disease; Parasitoids; Predation*

Parasitoids

There are many unpleasant ways of dying, but slowly being eaten alive must be near the top of the list: this is the fate of billions of insects that are attacked by parasitoids.

Parasitoids are insects (mostly wasps) of which females locate, but don't kill, their prey; they lay one egg or more on or in the body of their victim. The live prey now becomes a fresh (and in some cases, still developing) food store for the larval parasitoid. The end point is the host's death.

This way of life is no mere oddity. Almost 10% of known insect species are parasitoids, which means there are more parasitoids than species of mammals, birds, reptiles, amphibians and fish put together. They are also a vital weapon in the battle against insect pests.

Parasitoids often find their prey using chemical cues. For example, parasitoids that attack fruit flies are attracted to the smell of fermenting fruit; others are attracted by chemicals given off by plants as they are eaten by insects.

Normally only one larva can survive inside a host, so it's not a good idea to lay an egg in a host that's already parasitised. Parasitoids have therefore evolved the ability to discriminate between parasitised and unparasitised hosts, using chemical cues left by previously ovipositing females.

Hosts aren't entirely defenceless they have behavioural, physical and chemical protection. Even if an egg is laid inside their body, they have an immune response in which specialised blood cells surround and encapsulate the egg, leading to its suffocation or starvation. Parasitoids may counteract this by injecting a chemical that causes the blood cells to lose their stickiness; some even engage in germ warfare by injecting a virus along with the egg that interferes with the host's immune system.

see also...

Biocontrol of pests; Chemical ecology; Coevolution; Predation

Patchiness

Most, if not all, habitats are patchy at some level, whether it is islands in the sea, or mushrooms on a forest floor. A patchy environment consists of discrete patches of habitat that are distinguishable from their surroundings. One group of organisms that live in an extremely patchy environment are parasites – here individual hosts are patches.

The extent to which a habitat is considered patchy depends on the size and mobility of the species concerned. Even in habitats that aren't noticeably patchy, organisms often show what is known as distributional patchiness, in which individuals tend to clump together. The concept of a patchy world permeates current ecological thinking. For example, the idea of metapopulation dynamics is based on a patchy habitat with limited movement between patches. And in the 'patch dynamics' view, apparently stable communities (such as forests) are seen as consisting of a mosaic of patches, all at different stages of recovery from disturbances.

Coexistence of competing species,

and predators and their prey, is often enhanced in patchy habitats, as long as the weaker competitor or the prey can keep one step ahead. The classic example of this was an experiment (that was way ahead of its time) involving two species of mite, some oranges, rubber balls, and a jar of Vaseline. In a relatively homogeneous environment (oranges packed closely together) the predator soon found and ate all its prey, and then promptly starved to death. However, when the environment was made more patchy, by interspersing the oranges with rubber balls and making it difficult for the predator to walk between patches by using Vaseline barriers, it was found that predator and prey could coexist. Individual 'prey' could move freely launching themselves to safety on silken strands. At any one time this system consisted of a mosaic of unoccupied patches, patches containing prey only, and patches in which the predator had caught up with the prey.

see also...

Dispersal; Metapopulation; Species coexistence

Population

The study of populations is a central part of ecology. A population is simply a group of organisms of the same species living in a given area at a particular time. As a lot of populations don't have an obvious boundary it is often left up to the ecologist to make a pragmatic decision about the size of the area.

To make the figures more comparable between areas, ecologists often count the number of individuals in a given area or volume (the population density). Even this omits some information: the way in which individuals are distributed in space (e.g. clumped or random) can be important too, as it will affect the chances of interactions between them.

A balance of just four fundamental processes determines population size: birth, death, emigration and immigration. Ecologists have tended to focus on the first two, as they're often easier to measure, but it's recognised that movement into and out of populations is very important (especially in 'metapopulations'), and these processes are now being looked at more seriously.

Simply counting the number of individuals in an area hides a lot of important information. For example, all populations have an age structure (immature individuals, breeding adults, and post-reproductive stages), and consist of genetically different individuals. To take an extreme example, if all the members of a population are past the breeding age then its future is quite different from a population of the same size, but consisting of breeding individuals. The sex ratio can also be important, particularly in extremely rare species. The last 18 kakapo (the world's only nocturnal, flightless parrot) on mainland New Zealand might have stood a chance of recovery, if they weren't all males.

The study of populations has a number of very important applications, for example, in pest control, game management, the sustainable harvesting of animals and plants (e.g. fisheries), and the conservation of rare species.

see also...

Metapopulation; Population growth; Population regulation

Population growth

I once read that if the human population continued to grow at its present rate, in 200 hundred years' there would be a solid mass of people expanding into space at the speed of light. This isn't going to happen, but it shows the potential for population growth in all species.

Under 'ideal' conditions, population growth is determined by two factors: number of reproducing individuals, and number of offspring produced per individual that in turn go on to reproduce. So, even when the number of surviving offspring is constant, a population will continue to grow at an ever-faster rate because more individuals are constantly being added. This situation displays 'exponential' growth; not only does population size increase, it grows more quickly.

Populations in nature sometimes show exponential growth, for instance when a species is introduced into an area, but it isn't sustainable. Eventually population growth rate will either slow down and stabilise around its carrying capacity (the population size that's sustainable in a particular environment), or overshoot carrying capacity and crash ('boom and bust' dynamics), or anything between the two.

Population growth rate normally declines as the population increases beyond a certain size because there are not enough resources to sustain continued growth, individuals pollute their environment, are held in check by other species, or some combination of these. These 'density-dependent' processes act as a brake on population growth rate because their effects on birth and death rates are stronger at higher population densities.

The human population is accelerating at a rate that's even faster than exponential; it's doubling in size about every 35 years, which isn't sustainable. What is the carrying capacity of the human population? Pick a number between 10 and 1000 billion. It's not just that more individuals are being added to the global population; people are living longer and are becoming ever more resource hungry.

see also...

Density dependence; Population; Population regulation

Population regulation

Some populations show remarkable consistency in population size over long periods of time, others fluctuate quite considerably. In neither case, however, do populations go extinct nor, despite the enormous potential for population growth, do they show unlimited growth. For this to occur over extended periods suggests that something is regulating population size.

Populations can only be regulated by the action of negative feedback processes that allow a population to increase in size at low densities, but reduce population growth rate as density increases. In other words, populations can only be regulated by density-dependent processes. Sometimes the link between population size and density dependence might be rather loose, with density dependence only kicking in above a certain threshold density; acting intermittently; or progressively increasing in strength as density increases.

Although population regulation necessarily involves the operation of density-dependent processes, these don't necessarily lead to a population being regulated. Density dependence might, for example, be too weak to counteract the effects of density-independent factors.

Not all population regulation involves regulation around a single equilibrium population size, for example strong density-dependent processes can cause populations to undergo periodic cycles, or even produce chaos. In all cases, however, the key feature is that population regulation keeps the population within bounds, and what's more, it returns the population to within these bounds if perturbed away from them.

What about metapopulations? Here, local populations do go extinct fairly frequently, so how can these be said to be regulated? Although the population in each habitat patch obviously isn't regulated, regulation occurs at the regional metapopulation level.

see also...

Balance of nature; Chaos; Density dependence; Equilibrium; Population growth

Predation

Lions, wolves and crocodiles are the sorts of animal we think of as predators, but these are only one type of predator (broadly defined as an organism that eats all or part of another organism while that organism is still alive). In this sense, seed-eating birds, parasites, parasitoids and grazing rabbits are predators. The predator usually gains through the interaction, while the prey loses. If these effects translate to changes in population size then, in the jargon (see Species interactions, p. 90), predation is a form of (+, –) interaction. However, not all (+, –) interactions are due to predation.

There is selection pressure on the prey to avoid becoming a meal. Mechanisms include running away, chemical defences, camouflage, and living in large groups. If the prey has a penchant for fighting back, then predators have a tendency to go for the old, young or infirm, as these are generally easier to catch. As these categories of prey aren't currently making a contribution to the next generation, their removal will have little impact on the future size of the prey population. In addition, by removing individuals from a prey population, there will be fewer individuals left to compete for resources, so the survivors often grow more quickly and produce more offspring than they otherwise would have done. This concept is the basis of sustainable harvesting, where a proportion can be removed without affecting the overall population size.

Predation forms the basis of biological pest control and can be very important in maintaining species diversity. However, when general predators are introduced to a new regions, especially islands, they can wreak havoc. For example, the decimation of flightless birds on islands to which cats have been brought. Predators also form the basis of biological pest control.

see also...

Biocontrol of pests; Herbivory; Parasitism; Parasitoids; Predator-mediated coexistence; Species interactions

Predator-mediated coexistence

Important tools in the management of grassland nature reserves are sheep and other grazing animals. Stocked at appropriate densities at the appropriate time of year, their single-minded approach to grazing reduces the abundance of competitively dominant grasses, allowing the survival of less competitive species thereby maintaining, or even enhancing, plant species diversity. This is predator-mediated coexistence, the most famous example of which is the 'keystone' predator, the starfish *Pisaster*, on rocky shores. Predator here is used in its broadest sense to include parasites, parasitoids and herbivores.

There are two main ways in which a predator can enable competing species to coexist in situations where weaker competitors would otherwise be excluded. The predator preferentially (or only) eats the stronger competitor; or the predator has no preference but concentrates disproportionately on the most abundant prey species. If that changes, as long as the predator shows 'switching behaviour' it will shift its attention to whatever species becomes most abundant.

In each of these cases the predator has the potential to enhance the survival of the weaker or rarer prey. Whether this potential is realised depends on the degree of preference shown by the predator and the intensity of predation. A high predation pressure is more likely to lead to the loss of species rather than its maintenance. Conversely, if the preference for the dominant prey, or the intensity of predation, is too weak then it might not be sufficient to keep the dominant prey species in check.

Other important factors to bear in mind are traits possessed by the prey species. In plants, for example, how they respond to grazing and possession of herbivore defences are important. Under certain conditions, predator-mediated coexistence is an important mechanism competing species to coexist. It therefore plays a key role in the maintenance of species diversity.

see also...

Disturbance; Indirect effects; Keystone species; Predation

Primary production

While animals obtain energy by eating other organisms, green plants and some microbes obtain energy directly from the sun. They convert and store this energy in high-energy carbohydrates using CO_2 from the air. Some of this energy 'fixed' by photosynthesis is used by the plant for its own metabolism; the rest is converted into biomass which is then available for those organisms that feed on plants. Biomass is the mass of living material in a given area/volume; it also includes dead bits still attached to the organism, such as tree trunks.

The accumulation of new biomass by photosynthesising plants over a given time period (usually one year) is called net primary production. A related term, primary productivity, is the rate at which new biomass is generated per unit area/volume. A lot of new plant biomass is produced each year in the ocean (high primary production), because they're enormous; actual productivity (biomass production per unit area) is quite small.

Primary productivity varies widely. The most productive ecosystems include tropical rainforests and estuaries. On land the most productive ecosystems are generally nearer the Equator where it's warmer, while at any particular latitude productivity is broadly related to water availability. Nutrient availability can also be important in limiting primary productivity, which is why adding fertilisers increases crop yields. Nutrients, such as nitrates and iron, and light are the most important factors limiting productivity in the oceans.

Primary productivity is a key ecosystem process; all consumer organisms ultimately depend on it. Humans appropriate about 45% of the world's terrestrial primary production. Increasing atmospheric CO_2 concentration is likely to have a large impact on global productivity patterns. However, the interactions between carbon dioxide, temperature, moisture and nutrient levels are complex, so it's not yet clear what the precise effects will be.

see also...

Decomposition; Ecological energetics; Ecosystem; Limiting factors; Trophic level(s)

Rarity

Most species are naturally rare. Such intrinsic rarity doesn't mean they are threatened with extinction; that threat generally arises when rarity is *imposed* on species by human activities.

There are seven ways of being rare, because rarity can be categorised along three dimensions: local abundance, geographical distribution, and habitat specificity. (All three dimensions are on a continuous scale – the first two are often correlated – but for simplicity let's keep the either/or combinations.) There are therefore eight possible combinations of these traits; seven of which are rare along at least one dimension. Knowing the way in which a species is rare is a useful first step towards deciding how to conserve it.

The perception of rarity often depends on scale. For example, some insects considered 'rare' in the British Isles are quite common, often occupying a wider range of habitats, on the continent. It's just that they're on the northern limit of their range in Britain.

Rarity has a fourth dimension too: time. It makes a difference whether a population is relatively constant, steadily declining, or plummeting. In recognition of this, The World Conservation Union (IUCN) now considers, among other things, the rates of decline in population size when assessing the threat of extinction. For example, a species that has declined by at least 80% over the last 10 years or three generations is labelled as 'critically endangered'.

Are there any general traits that make some species more prone to rarity? Maybe; but it's probably too broad a question to be useful. For some traits (e.g. dispersal ability) it's not clear whether they are a cause or a consequence of rarity.

The causes of imposed rarity are many and idiosyncratic, and are best tackled on a case-by-case basis.

> *see also...*
>
> *Biodiversity; Conservation; Minimum viable population*

Resource partitioning

Imagine two species of seed-eating finch that live in the same habitat. Seeds are in short supply and so there is competition for them. Despite competition, the species will be able to coexist provided that there is some degree of resource partitioning, in this case, each species concentrates on different size ranges of seeds. Resource partitioning works because each species inhibits its own growth more than its competitor, which is one of the necessary conditions for competing species to coexist, as it prevents either species from becoming common enough to exclude the other. Exactly how much overlap in resource use is allowed before one species excludes the other will vary between species, as it is dependent on many factors.

The resource may be partitioned in terms of the habitats used, or by some property of the resource itself, such as food type or size. Some ecologists argue that time is another dimension along which resources can be partitioned (temporal resource partitioning), although this is a bit trickier to envisage. Unless the species also use different type of resources, or the resource regenerates quickly enough, temporal resource partitioning is unlikely to be effective, as what's eaten by the earlier species isn't available for the later species.

Resource partitioning is thought to be a common means by which species undergoing exploitation competition (see Intraspecific competition, p. 51) are able to coexist. However, the reverse isn't necessarily true: simply observing resource partitioning in nature doesn't mean that the species are competing. It could be that the species have competed in the past, and resource partitioning has led to them not competing any more, or that competition led to the extinction of those species with too high an overlap, so we're left with those species that are different enough to coexist, or the resource isn't limited, in which case overlap in resource use is irrelevant.

see also...

Character displacement;
Interspecific competition; Niche;
Species coexistence

Restoration ecology

In recent years it has become common practice to restore damaged and degraded ecosystems. There are basically four options:

★ put back exactly what was there before (restoration);
★ return an ecosystem to something that resembles what was there before (rehabilitation);
★ turn it into another ecosystem (replacement);
★ leave it alone and let ecological succession undergo its course (neglect).

Of the techniques used, neglect can be the best ecological (and economic) solution. For example, following damage to coastal habitats by oil spills the best course of action is often inaction. The millions (or even billions) spent on restoring an oil-impacted shoreline is often ineffectual and at worst, the restoration attempt might result in more ecological damage than that caused by the original spill. How do we know if the restoration is successful? We can't use currently existing 'natural' ecosystems as our benchmark, as there are few, if any

left. Human impacts on some ecosystems go back thousands of years, others are more recent, but all have increased in intensity over time. So the question of how far back in time we should go to choose our benchmark ecosystem is essentially a value judgement. In North America, the aim is often to restore ecosystems to the state they were in before the arrival of European settlers (even though these had already been affected by the actions of indigenous tribes). In Europe, there isn't such a conveniently arbitrary starting point, and the reference conditions chosen are often those prevailing before World War II, and agricultural intensification.

The bits of ecosystem we restore are often small and isolated, which means that they are rarely self-sustaining and require continued management. Restoration is a valuable conservation tool, but it's only the first step.

see also...

Conservation; Habitat management; Habitat (re)creation; Historical ecology; Succession

Rivers

Only 0.0001% of all the water on the planet is found in rivers. In fact, at any given time there is ten times more water present in the atmosphere than is found in all the rivers of the world.

However, these bald facts belie the importance of rivers in terms of the global water cycle, and their importance to terrestrial organisms, including humans. Rivers drain a large proportion of the Earth's land surface, and have a very high throughput of water (discharging over 40,000 km^3 of it into the sea each year).

Rivers are unusual habitats in that they are dominated by a strong, one-way flow of water, downhill. This flow erodes, transports, and eventually deposits material (the world's rivers carry 15–20 billion tonnes of sediment to the sea each year). Turbulence and flow also directly and indirectly affect the lives of most river organisms.

Freshwater habitats in general are tougher places than the sea; largely due to their relatively small size, they are subjected to greater temperature fluctuations, their habitats are often physically separated from each other; and they can dry up or freeze. The organisms also have to cope with maintaining the salt balance in their bodies.

To understand river ecology fully, ecologists have to study the catchment area; it's from here that the river gets its water, its chemicals, and its nutrients. For example, in small rivers flowing through temperate forests, few photosynthesising plants can grow in the shade of trees overhanging the channel. They therefore depend on dead leaves falling from the trees to supply the bulk of the nutrients to the system. Trees are also important in that they reduce the amount of water entering the river; this explains why flooding increases dramatically after deforestation. Because of the flow of water and nutrients, downstream communities are very dependent on what happens upstream, reinforcing the view that the catchment is the key to understandng river ecology.

see also...

Lakes

Savannah

Tropical savannahs cover over half of Africa, occupying two broad belts either side of the Equator. They also occur in Australia, Brazil and southern Asia; all hot areas with highly seasonal rainfall (the driest three months may each have less than 5 cm of rain).

Savannahs consist of a continuous cover of grass with a few scattered trees. When we think of savannahs like the Serengeti we probably imagine vast herds of grazing ungulates (hoofed animals) on their mass migrations to avoid the dry season.

Leaving climate aside, fire and grazing are the dominant influences on savannahs, and they interact in complex ways. Both grazing and fire generally favour grasses rather than trees, as their growing points (meristems) are at or below ground level, and are therefore protected. In the absence of either factor, tree cover increases. Grazing of grasses also reduces the fuel load, reducing fire frequency, thereby increasing the abundance of woody plants.

The effect of herbivores in areas of African savannah with few fires was dramatically shown following their infection by the rinderpest virus, caught from domestic cattle. Occasional epidemics over the last 100 years periodically decimated ungulate populations, thereby reducing grazing pressure, with the result that narrow windows of opportunity were created for acacia tree seedlings to establish and grow into adult trees. The result is that in some areas acacia trees are virtually all the same age. Conversely, in areas more prone to fires, as the herbivore populations recovered, tree seedlings became established as increased grazing reduced fire frequencies.

We have a lot to learn about savannahs. What appears to be a simple habitat is a shifting mosaic of transitional states between grass and tree dominance, resulting from complex interactions between trees, grasses, climate, grazing, fire and increasingly varied human impacts.

see also...

Biomes; Grasslands

Scale in ecology

A lot of ecological processes operate over much larger (and smaller) spatial and temporal scales than our usual frames of reference. Ecology deals with a spatial scale ranging from microscopic to global, and a temporal scale from seconds to millennia.

In contrast, the majority of ecological studies last less than five years and the study area is rarely more than 10 m^2. This is important, because there's no logical reason to suppose that the processes operating over the spatial and temporal scales of most ecological studies are still important over larger spatial and temporal scales.

It has been said that most ecology is like trying to reconstruct a film, 'from a few consecutive frames of one film or from single frames of many films that we hope are similar' (Wiens *et al*, 1986). The upshot of this is that you can't fully understand ecology without an appreciation of scale. Freshwater ecologists have recognised this for some time, they realise that you can't fully understand the ecology of rivers without considering processes operating at the scale of the catchment area. Hence the increasing number of long-term studies, which are providing a valuable perspective on various ecological processes.

The size of the organisms ecologists study vary from the microscopic, to blue whales and giant sequoias, and size has important ecological implications. For example, generation time, abundance and metabolic rate are all correlated with size. Fishes can cut through water with the merest flick of the tail, whereas for micro-organisms moving in water is like moving through treacle. As size changes, so too does the relevant time scale. What seems like occasional ecological 'disasters' to us, are regular occurrences to trees that live for hundreds of years.

The effects of scale can't be ignored, and choosing the appropriate scale of study is one of the most important decisions for an ecologist.

see also...

Landscape ecology; Macroecology

Semantics

emantics has played a big role in many ecological controversies. For example, community ecology has been described as, possibly, being unique in science in that it lacks, 'a consensus definition of the entity [i.e. the community] with which it is principally concerned' (Giller and Gee, 1987). The controversy about the link between density dependence and population regulation was largely resolved when it was realised that the protagonists were arguing about different concepts (population regulation and population limitation) that were being used synonymously.

Ecology is often accused of using inexact language. It's been said that one reason why there is little consensus in terminology is that many ecologists, 'pay little attention to those outside their immediate circle…' preferring, like Lewis Carroll's Humpty Dumpty, 'to make words mean just what they wish them to mean.' (McIntosh, 1995). The implication is that sometimes ecologists can be lazy, they vaguely know what a term means and use their version of it without checking the original definition closely.

Sometimes a single term has a variety of meanings. For example, stability has aspects of resistance, resilience and persistence, and until this was clarified ecologists were often arguing because they were using the same term, but in a different sense. For example, the term niche has multiple, and quite different, meanings. To some biologists symbiosis is equated with mutualism, to others the terms are quite different. Sometimes biodiversity is taken to include ecosystem functioning, sometimes not.

To be fair to ecologists, a major part of the problem with definitions in ecology is that many of the objects or concepts are not easy to define. The definition of community is vague because communities are, by their very nature, vague. Having said that, even if all the terms in ecology were unambiguously defined, ecologists would still find plenty to argue about.

see also...

Chaos; Community; Guilds; Niche; Rarity; Symbiosis

Species/area relationship

One of the most widespread patterns in ecology is that the bigger the area the more species it contains. In fact the pattern is more specific than that: if you plot a graph of number of species against area you'd find that the rate at which new species accumulate decreases with increasing area in a fairly predictable way. A broad prediction arising from this is that if you destroy 90% of an area you will lose about half of the species. This has been used to argue that a single large nature reserve will support more species than several smaller reserves that add up to the same total area as the large one.

The relationship between area and number of species may seem obvious, but *why* would you find more species in a larger area? Three prominent explanations have been put forward. The first – island biogeography theory – says that extinction rates will be lower on larger islands because they can support higher population sizes, and larger populations are less likely to go extinct. Larger islands contain more species than smaller ones.

The second idea is that larger areas contain a greater diversity of habitats, therefore more species. This sounds plausible, but what exactly is meant by habitat diversity, and how can you quantify it in an ecologically meaningful way?

The third idea is that the relationship is simply a statistical artefact. As the sample area increases you will soon find all the commoner species; as it increases further the number of new species found will decrease as the only species left to find are increasingly rarer ones.

Which of these theories is best? Evidence has been found to support all three hypotheses and indeed, they aren't necessarily mutually exclusive, and different mechanisms are likely to be important at different spatial scales and for different communities.

see also...

Island biogeography theory;
Macroecology

Species coexistence

How can competing species coexist? Why doesn't the stronger competitor eliminate the weaker? There are two possible outcomes of competition between two species, either one eliminates the other (competitive exclusion), or they both coexist. Sometimes both occur, though on different spatial scales. One species of carrion fly might eliminate another at the scale of the decomposing mouse, but on a regional scale the species can coexist if the weaker competitor is the better disperser. Two species of barnacle can coexist on the same rocky shore, although one species excludes the other from the lower shore.

The key thing about stable coexistence is that the competitive abilities of the species must be inversely correlated with their relative frequency: the more abundant a species becomes the more it suffers from intra- (rather than inter-) specific competition.

The mechanism of coexistence that has dominated ecological thinking is Resource partitioning (see p. 82), but more recently other mechanisms have been put forward. The action of predators or disturbances can keep some species at levels at which competition isn't important, or their effects fall disproportionately on the stronger competitor (see Disturbance, p. 24, and Predator-mediated coexistence, p. 79).

Sometimes the trend towards competitive exclusion is halted or reversed by sufficiently frequent changes in environmental conditions. This is the explanation put forward as an answer to the 'paradox of the plankton' (i.e. how so many species of plankton coexist in an apparently uniform habitat when they all require the same handful of limited nutrients).

In patchy habitats coexistence can be promoted by a number of means. For example, if a superior competitor has a clumped distribution this leaves lots of unoccupied patches for the weaker competitor. There is also the possibility that apparently coexisting species are simply on a long-term trend to competitive exclusion.

see also...

Character displacement; Disturbance; Interspecific competition; Niche; Predator-mediated coexistence; Resource partitioning

Species interactions

Understanding interactions between species is arguably the most fundamental part of ecology.

One of the easiest ways to classify interactions between species is by measuring the effects each species has on the other; these can be beneficial, detrimental or neutral (no effect), signified by +, − and 0 signs respectively. This leads to five interaction types: interspecific competition (symbolised as (− , −)), amensalism (− , 0), commensalism (0 , +), contramensalism (+ , −) and mutualism (+ , +).

Each effect can be further divided, based on the mechanism involved. Contramensalism can arise from, among other things, trophic (feeding) interactions like predation and non-trophic interactions such as mimicry. Interspecific competition can arise when two species use the same limited resource, when they share a common predator, or when they eat each other.

Things start to get complicated now. What effect are we measuring? There could be effects on population (e.g.

population growth rate, equilibrium population size), or effects on individuals (e.g. growth rate, size, reproductive rate). While both levels are valid, effects on individuals (e.g. being killed) aren't always translated to effects on the population.

Even if we stick to effects on population size, is it better to measure the total impact of one population on another, or is it more useful to measure the per capita impact? It makes a difference.

Furthermore, interactions aren't fixed. The type of interaction often depends on environmental conditions, and on other species present (via 'indirect effects'). For example, the interaction between two species of fruit fly and a species of fungus in a simplified laboratory system was found to be highly variable, being dependent on the precise environmental conditions. In fact, all of the interaction types listed were found.

see also...

Herbivory; Indirect effects; Interspecific competition; Mutualism; Parasitism; Predation

Species introductions

Since Europeans first arrived in the United States, they have brought with them over 50,000 species. The economic losses, human health impacts, and costs of controlling introduced species in the USA exceeds US$138 billion a year. Take the black rat, for example. Each rat eats about US$15 worth of food each year. With an estimated 1.25 billion rats in the USA, you have annual losses of US$19 billion.

Introduced species don't just have economic costs. They can cause immense ecological harm by causing disease, eating or out-competing native species, and altering habitats. They are one of the most important threats to biodiversity, and are leading to homogenisation of the world's biota.

Oceanic islands, because of their isolation, are home to a huge diversity of endemic species (species found nowhere else), and are particularly vulnerable to the effects of introduced species. On the island of Guam, for instance, only three of the original 13 forest bird species survive; half the reptile species have gone; and the only surviving native mammal (a fruit bat) hasn't bred in over a decade. The blame for most of these extinctions has been put on the brown treesnake, which invaded the island in army vehicles and aircraft after World War II. How could one species wreak such havoc?

First, as on other oceanic islands, Guam's vertebrates evolved largely in the absence of predators, and exhibit 'island tameness': they don't run away when approached and are easy prey for introduced predators.

Second, the treesnake isn't the only invader; the island is now home to introduced rodents, curious skinks and mutilating geckos. This abundant source of introduced prey helped the snake, with its catholic food tastes, to attain very high densities, putting an intolerable pressure on the native fauna became unsustainable, and several species went extinct. Having eaten much native fauna to extinction, the snake now survives by eating other introduced fauna.

> ### see also...
> *Biocontrol of pests; Indirect effects*

Succession

Ecological succession is one of the central organising ideas in ecology. Succession is the broadly directional replacement of one community by another, following disturbance of a site. Because succession can take places over centuries, the process isn't completely amenable to experiments. Ecologists therefore often substitute space for time: sites of different ages are interpreted as different time points for one site, which may not always be valid.

Succession is a complex process involving many factors, but still results in broadly recognisable, predictable patterns of recovery from a disturbance. Among all individual site-specific successions, some generalities emerge, based on allocation and trade-offs of resources within species (see Life history strategies, p. 58). Species with good dispersal abilities and fast growth rates tend to dominate the early stage. However, as these species are not usually good competitors, they will eventually be replaced by later arriving, slower growing, more competitive species. As succession proceeds, the trend is towards species that grow more slowly, live longer, are bigger and more shade tolerant, and are poor dispersers.

Although differences in life history traits alone are theoretically sufficient to explain the general pattern of succession, interactions between species also play a role, at least in affecting the rate. Early-arriving species can have a range of net effects on later arriving species: they may inhibit them, facilitate them, have no effect, or anything in between.

An understanding of the principles underlying the process of succession is important when ecologists attempt to restore post-industrial landscapes and when managing early or mid-successional habitats for conservation.

see also...

Disturbance; Life history strategies; Species interactions; Succession – primary, Succession – secondary

Succession – primary

The eruption of Mount St Helens in Washington State in 1980 killed virtually everything over several square kilometres. It melted nearby glaciers, turning local rivers into surging mudflows; the ash spewed from the crater covered a third of the state to a depth of 4 cm; and it flattened or snapped trees in half up to 20 km away. Such major disturbances wipe the slate clean: there are no animals, no plants, and no soil. Soon, however, organisms start to arrive, and the process of primary succession begins.

In the early stages of primary succession factors influencing colonisation are important, such as the closeness of seed sources. The main problem faced by plants once they've arrived is that there are few or no nutrients in the soil, or even any soil. Lichens can be important here as their ability to disintegrate rocks by physical and chemical means helps in the formation of soil. Some species can survive nutrient-poor conditions because they have symbiotic bacteria living in their roots; these can 'fix' atmospheric nitrogen, making it available in a form usable to plants.

Meanwhile, on quiet beaches around the world, small plants, tolerant of the salty air and the arid conditions, are being covered with wind-blown sand. These plants are important as they add organic matter and nutrients to the soil, which also improves the sand's water retention. Succession really gets going when species like marram grass arrive. Their extensive root system binds the dunes together and improves the soil conditions, enabling other species to survive. In other words, marram facilitates (though not in any altruistic sense) the establishment of other species. Such general facilitation is important in primary succession.

Having paved the way for other species, marram is then eliminated by them. Recent work suggests that pathogens of later arriving species may play an important part in the elimination of early-successional species like marram. Conditions are now suitable for many more species and a grassland community develops.

see also...

Succession; Succession – secondary

93

Succession – secondary

Secondary succession occurs when the vegetation has been removed from an area, but the mature soil remains essentially intact. A simple explanation proposed for secondary succession is that it depends on the 'initial floristic composition' of an area. Under this scenario succession is simply a consequence of the different growth rates of the species already present in the seed bank or as tubers, so in forested regions the community returns to forest mainly because trees take longest to grow.

Thus, colonisation processes tend to be less important in secondary succession, compared with primary succession. Also, because the soil is well established, facilitation is thought to be less important in secondary succession.

Herbivory can affect the rate of vegetational succession. Grazing by rabbits, for instance, slows down the change from grassland to scrub, although there are fewer examples of native herbivores affecting the rate of succession.

Is there a single climatic 'climax community' (end point) for a given region? No. Assuming there is any such thing as a climax community (which some ecologists doubt), there will be several in a region, depending on local circumstances such as soil conditions and the frequency of disturbance. For example, within the general pattern of succession from lakes to semi-aquatic reedswamp to dry land, there is a lot of complexity. Pollen analysis indicates that sometimes the succession seems to go in the 'wrong direction', depending on the prevailing environmental conditions. For example, dry woodland quite frequently reverted back to wetter *Sphagnum* moss communities.

Does succession ever stop? No, it just gets slower when long-lived species, like trees, arrive. Even in apparently stable forests, however, local disturbances will create gaps and the process of succession will begin again. Succession also occurs on much larger time scales: the vegetation of the Northern hemisphere is still recovering from the last Ice Age.

see also...

Disturbance; Fire; Succession; Succession – primary

Symbiosis

iftia, a metre-long tube worm is an important component of the hydrothermal vent communities, found in mid-ocean ridges where the ocean floor is spreading apart. *Riftia* has neither a mouth nor any digestive system. So how does it survive? The answer lies inside the organ that makes up most of its body. This organ contains vast numbers of bacteria that use sulphur-containing chemicals as their energy source (most 'autotrophs', such as plants, use energy captured from sunlight by the process of photosynthesis). These bacteria provide *Riftia* with its carbohydrates; in return *Riftia* provides the bacteria with CO_2, oxygen and hydrogen sulphide, raw materials for their 'chemosynthesis'.

The relationship between *Riftia* and the bacteria is an example of symbiosis, in which two species live in close physical association with each other. In this case the bacteria are 'endosymbionts' as they live completely inside their 'host'. Symbioses aren't necessarily mutually beneficial. The commonest form of symbiosis is where one organism parasitises another.

There is a close link between parasitic and mutualistic symbioses. For example, virtually all plant species have fungi living inside their aboveground tissues, indeed trees contain dozens of them. These fungi, many of which are regarded as beneficial to their plant hosts, probably evolved from an initially parasitic relationship.

The tube worms, with their energy-fixing bacteria, are analogous to green plants with their energy-fixing organelles (chloroplasts) contained within their cells and used for photosynthesis. The analogy is even closer than at first sight, for it is now believed that chloroplasts were once free-living bacteria that have become endosymbiotic with plants. In addition, all multicellular organisms contain other ex-bacteria in their cells, in the form of mitochondria (organelles that convert energy stored in carbohydrates into energy directly usable by cells).

> ### see also...
> *Coevolution; Mutualism; Parasitism*

Temperate forests

The most familiar type of temperate forest, at least to inhabitants of the Northern hemisphere, consists mainly of deciduous trees that shed their leaves in autumn.

Temperate deciduous forests occur in areas with quite large seasonal temperature fluctuations – cool or cold winters and warm summers – and with quite high rainfall throughout the year. In terms of appearance, this biome probably shows the greatest changes throughout the seasons. Most of the plants are dormant throughout the winter, with the early-flowering ground plants lying dormant as bulbs or other underground storage organs. This enables them to get a head start when spring arrives, before the tree canopy closes over them.

Woods are three-dimensional habitats, with several distinct layers; the total leaf area is several times that of the ground surface. In summer, the dense canopy shade means that little sunlight reaches the ground. Some shade-tolerant ground flora is still present, often in the lighter parts of the wood.

In autumn the trees withdraw as much minerals and nutrients from the leaves as they can, causing them to change colour before they fall. The fallen leaves provide a large resource for the soil decomposer community.

Forests are dynamic systems over a wide range of temporal and spatial scales. For example, the main tree species in the temperate forests of north east America are temporary associations, rather than a highly integrated community. Since the ice caps retreated after the last glaciation tree species have been moving northwards independently of each other and it's only recently, historically speaking, that their paths have crossed to form the forests we see today. The dynamic nature of forests is also seen at a regional level; forests are not so much a green blanket as a patchwork quilt. Disturbances are frequent, with the result that different areas are at different stages of recovery.

see also...

Coniferous forests (taiga)

Top-down/bottom-up

Forty years ago three ecologists asked a question, why is the world green? In other words, why isn't all that lush green terrestrial vegetation devoured by herbivores?

The answer they proposed was this. The numbers of herbivores are held in check by the effects of natural enemies, e.g. predators and parasitoids, thus only a fraction (about a fifth) of the vegetation is consumed. The numbers of herbivores are therefore limited by 'top-down' forces from the trophic level above. The logical inference from this is that the bottom trophic level, plants, are not limited by herbivores (as there aren't enough of them), and so plant biomass increases until it becomes limited by the supply of nutrients (bottom-up limitation). Similarly, the predators that feed on the herbivores are limited by competition for a limited supply of herbivores. This argument can be extended to include more or fewer trophic levels. A nice idea, but is it true?

Leaving aside the criticism that the idea of trophic levels is too simplistic, there are two other reasons to explain why the world is green. First, plants rarely get eaten because they are protected by an array of physical and chemical defences; and second, plants are a rather poor-quality food, particularly in terms of their low nitrogen content.

Top-down forces are clearly important in some communities, as illustrated by the phenomenon of trophic cascades. However, at the most basic level bottom-up limitation must be the most important, removing all the predators from an ecosystem will have serious consequences, but not as serious as removing all the plants as here the whole system will collapse. An important question is how far up or down the food chain the effects of nutrients and predators are felt.

Current opinion is that a top-down/bottom-up dichotomy is too simplistic. Both have a role to play in most ecosystems, and the question has shifted to how their effects interact in different systems.

---

see also...

Herbivory; Primary production;
Trophic cascade(s); Trophic level(s)

Trophic cascade(s)

Sea otters eat sea urchins eat kelp. Until overhunting the otter nearly caused its extinction, which was also unfortunate for the kelp as it meant that there was nothing to keep the numbers of sea urchins in check. This resulted in the kelp being overgrazed and eliminated from some areas, along with all the other species dependent on it. This is a trophic (feeding) cascade.

When otter numbers increased after hunting stopped, the kelp started to return. More recently, however, it's been suggested that overfishing has caused a reduction in the number of seals and sea lions, the preferred prey of killer whales. Whales have now begun to hunt otters, causing a decline in their population.

Trophic cascades are a form of indirect mutualism. More formally, trophic cascades are 'reciprocal predator–prey effects that alter the abundance, biomass or productivity of a population, community or trophic level across more than one link in a food web' (Pace *et al*, 1999).

The study of trophic cascades has important practical applications. For example, the principle has been applied to the restoration of shallow eutrophic lakes. These lakes suffer from algal blooms resulting from excessive input of plant nutrients. This has severe effects on the whole lake community, resulting in the loss of many species. One way of restoring eutrophic lakes (combined with reducing nutrient input) is to remove the fish that eat the zooplankton that feed on the algae. Or introduce fish that feed on the fish that feed on the zooplankton. Either way, the result is an increase in zooplankton numbers, helping to keep algae under control.

Finally, there are potentially serious implications arising from our over-exploitation of (mostly predatory) marine fish stocks. As the predators are removed their prey populations increase, so we start to fish these.

see also...

Community – alternative stable states; Food web; Indirect effects; Keystone species; Mutualism; Top-down/bottom-up; Trophic level(s)

Trophic level(s)

Grouping organisms into broad categories – trophic levels – based on their position in the 'food chain' has been seen as a useful simplification when it comes to understanding ecosystem structure and function in terms of energy flow.

At the bottom of the food chain are the primary producers (mainly plants) which are eaten by herbivores (primary consumers), which are eaten by predators (secondary consumers), which are in turn eaten by bigger predators (tertiary consumers), and so on. In terms of biomass, when the trophic levels are stacked like Lego® bricks, with the size of the bricks proportional to the amount of biomass represented, the result is a pyramid shape with primary producers on the bottom. Sometimes the pyramids are inverted, with a broad herbivore trophic level perched precariously on top of a small primary producer level, implying that the producer biomass is actually less than the herbivore biomass it is supporting. Although at any particular moment in time this might be true, this is a static view and ignores the dynamics of the system. By reproducing at a very high rate the producer biomass is continually being replenished at a much faster rate than the consumer biomass; it's just that it's being consumed at a high rate too. If trophic levels are viewed in terms of energy input then, because of inefficiencies in energy transfer between levels, there must always be a pyramid shape (see Ecological energetics p. 25).

The idea of trophic levels has come in for criticism. Are they simply a jargonistic way of stating the obvious? Is it too broad a concept to be useful? Where do omnivores and decomposers go? And what about carnivorous plants? Adult red grouse eat young heather shoots, while their offspring eat insects, placing individuals of the same species in different trophic levels. Because of these difficulties some ecologists argue that the idea of trophic levels serves no useful purpose and should be abandoned. They argue that food webs are a more useful way of representing energy flow and transformation in ecosystems.

see also...

Ecological energetics; Ecosystem; Food web; Primary production

Tropical rainforests

People can be disappointed the first time they enter a tropical rainforest. It's rather gloomy underneath the tight canopy of tall trees. As a result there isn't much undergrowth, so hacking through the 'jungle' with a machete is not necessary.

You see very few animals: most live in the tree canopy, where the flowers, fruit and leaves are. At ground level its hard to believe this is the most species-rich habitat on the planet, containing about 40% of the world's plant species, with dozens (hundreds in some areas) of tree species per hectare.

Tropical rainforests are found in the warm, moist lowlands within 10° either side of the Equator, in Brazil, central Africa and South-east Asia. There are thundershowers almost every afternoon, resulting in over 2 metres of rainfall a year. The very old soils contain few nutrients but, because the forest is extremely efficient at recycling nutrients, they support one of the world's most productive habitats.

The continual rain of shed leaves returns nutrients to the soil, which then decompose rapidly in the warm, moist conditions, and are quickly taken back up by the tree roots, before the rain washes them away. Two factors aid efficient nutrient recycling: most roots are sheathed with mutualistic mycorrhizal fungi, which enhance nutrient uptake; and the roots are concentrated in the top 20 cm of soil, so they can absorb nutrients as soon as they become available. Tall trees with shallow roots have buttresses on the trunks that act like guy ropes to help stabilise them; they also funnel water and nutrients running down the trunk to the roots.

Crops grown on areas by 'slash-and-burn' farmers aren't as good at holding onto nutrients as the forest they've replaced; nutrients are also removed from the system when the crop is harvested. So, after a couple of years, the soil is too depleted for crops to grow, and the farmers move on to burn down another area.

see also...

Global environmental change;
Latitudinal diversity gradient;
Mutualism

Tundra

In the land of the midnight sun, weather isn't high on the list of interesting conversational gambits. In the tundra – treeless expanses found north of the taiga – temperatures are way below freezing for most of the year, and even the warmest months of the brief summer, temperatures average only around 5°C. In the higher latitudes it is too cold for the air to hold much moisture, so it doesn't actually snow very much; it's just that it when it does it hangs around for a long time. For most of the year usable water is in short supply; it's either locked up as snow above ground or as ice below ground, and the desiccating winds add to the water shortage. The tundra is basically an extremely cold desert.

A metre below the surface the ground is permanently frozen (permafrost). This impedes drainage and so, even when the top layer of soil thaws out in the summer, it stays wet, often forming shallow pools. In the cold, waterlogged soils decomposition rates are slow, and a layer of partially decomposed organic matter often carpets the ground, accumulating as peat. Plant nutrients, particularly nitrogen, are therefore in short supply.

Plants found in the tundra are short, which at least allows them to gain some protection from the harsh winter conditions under the thin blanket of snow. Tussocky grasses and sedges dominate the landscape, with a few dwarf shrubs clinging to slightly warmer pockets. The leaves and stems of several Arctic plants are an attractive purple colour, helping them to warm up, even enabling some of them to absorb enough energy to begin growth in the spring while still under the snow. Annual plants are rare; the growing season is too short for them to complete their life cycle. In the harshest tundra conditions only mosses and lichens are found; these can endure snow cover for several years.

Over 100 species of bird breed in the tundra but most of them migrate south for the winter. The characteristic mammals are several species of small rodents and the much larger caribou and musk ox.

see also...

Coniferous forests (taiga)

Further Reading

General textbooks

T. F. H. Allen, S. R. Carpenter, S. I. Dodson, A. R. Ives, J. F. Kitchell, N. E. Langston, R. L. Leanne and M. G. Turner, *Ecology* (London, Oxford University Press, 1998)

C. J. Krebs, *Ecology*, 4th edn (Benjamin/Cummings, 1994)

J. R. Krebs and N. B. Davies (eds.) *Behavioural Ecology: An Evolutionary Approach*, 4th edn (Oxford, Blackwell Science, 1997)

P. E. Odum, *Ecology: A Bridge between Science and Society* (Sunderland, MA, Sinauer Associates, 1997)

R. E. Ricklefs and G. L. Miller, *Ecology*, 4th edn (Basingstoke, W.H. Freeman & Company, 2000)

P. Stiling, *Ecology: Theories and Applications*, 3rd edn (Prentice Hall, 1999)

C. R. Townsend, J. L. Harper, and M. Begon, *Essentials of Ecology* (Oxford, Blackwell Science, 2000)

Applied

A. Beeby, *Applying Ecology* (Chapman & Hall, 1993)

O. L. Gilbert and P. Anderson, *Habitat Creation and Repair* (London, Oxford University Press, 1998)

G. K. Meffe and C. R. Carroll, *Principles of Conservation Biology*, 2nd edn. (Sunderland, MA, Sinauer Associates, 1997)

E. I. Newman, *Applied Ecology* (Oxford, Blackwell Science, 1993)

W. J. Sutherland (ed.), *Conservation Science and Action* (Oxford, Blackwell Science, 1998)

W. J. Sutherland and D. A. Hill (eds.), *Managing Habitats for Conservation* (Cambridge, Cambridge University Press, 1995)

C. H. Walker, S. P. Hopkin, R. M. Sibley and D. B. Peakall, *Principles of Ecotoxicology* (London, Taylor & Francis, 1996)

Aquatic

R. S. K. Barnes and R. N. Hughes, *Introduction to Marine Ecology*, 3rd edn (Oxford, Blackwell Science, 1999)

M. Dobson and C. Frid, *Ecology of Aquatic Systems* (London, Longman Higher Education, 1998)

B. Moss, *Ecology of Freshwaters: Man and Medium, Past to Future*, 3rd edn (Oxford, Blackwell Science, 1998)

Microbial

R. M. Atlas and R. Bartha, *Microbial Ecology: Fundamentals and Applications* (Benjamin/Cummings, 1998)

General Reading

C. Darwin, *The Origin of Species* (Harmondsworth, Penguin Books, 1982)

J. Lovelock, *The Ages of Gaia* (London, Oxford University Press, 1995)

D. Quammen, *The Song of the Dodo. Island Biogeography in an Age of Extinction* (London, Pimlico, 1997)

E. O. Wilson, *The Diversity of Life* (Harmondsworth, Penguin Books, 1992)

General References

R. P. McIntosh, 'H.A. Gleason's "Individualistic Concept" and Theory of Animal Communities: A Continuity Controversy', *Biological Reviews of the Cambridge Philosophical Society*, (Cambridge, Cambridge University Press, 1995), Vol. 70, pp.317–357

R. M. May, 'Unanswered questions in ecology', *Philosophical Transactions of the Royal Society of London* (London, The Royal Society, 1999), vol. B354, pp.1951–1959

M. L. Pace, J. J. Cole, S. R. Carpenter and J. F. Kitchell, 'Trophic cascades revealed in diverse ecosystems', *Trends in Ecology and Evolution* (London, Elsevier Science, 1999), vol. 14, pp. 483–488

M. E. Power and L. S. Mills, 'The keystone cops meet in Hilo', *Trends in Ecology and Evolution* (London, Elsevier Science, 1999), vol. 10, pp. 182–184

O. Rackham, 'Implications of historical ecology for conservation', *Conservation Science and Action*, ed. W. J. Sutherland (Oxford, Blackwell Science, 1998)

R. B. Root, 'The niche exploitation pattern of the blue-gray gnatcatcher', *Ecological Monographs* (Ecological Society of America, 1967), vol. 37, pp. 317–350

I. F. Spellerberg, *Conservation Biology* (London, Longman Higher Education, 1996)

W. J. Sutherland (ed.), *Conservation, Science and Action* (Oxford, Blackwell Science, 1998)

J. A. Wiens, J. F. Addicott, J. J. Case and J. Diamond (eds), 'Overview: The Importance of Spatial and Temporal Scale in Ecological Investigations', *Community Ecology* (New York, Harper & Row, 1986)

Also available in the series

TY 101 Key Ideas: Astronomy	Jim Breithaupt	0 340 78214 5
TY 101 Key Ideas: Buddhism	Mel Thompson	0 340 78028 2
TY 101 Key Ideas: Evolution	Morton Jenkins	0 340 78210 2
TY 101 Key Ideas: Existentialism	George Myerson	0 340 78152 1
TY 101 Key Ideas: Genetics	Morton Jenkins	0 340 78211 0
TY 101 Key Ideas: Philosophy	Paul Oliver	0 340 78029 0
TY 101 Key Ideas: Psychology	Dave Robinson	0 340 78155 6